Y0-BRJ-215

The Practical Handbook of

TRANSPORTATION CONTRACTING AND RATE NEGOTIATIONS

First Edition

Practical Handbook of

TRANSPORTATION CONTRACTING AND RATE NEGOTIATIONS

Colin Barrett

Published by THE TRAFFIC SERVICE CORPORATION
Washington • New York • Chicago • Westport
Boston • Atlanta • Palo Alto

1325 G Street, N.W., Washington, D.C. 20005

First Edition

Library of Congress Catalog Card Number 86-050988

ISBN 0-87408-037-1

Printed in the United States of America

Produced by Stephen R. Hunter

Table of Contents

To the late Pete Hamm, who gave me opportunities, help and friendship.

To Steve Hunter, my friend—and his wife Ann.

To Mal Newbourne, also my friend—and his wife Edith.

To the late Joe Scheleen, again.

To my parents, and my sister.

And to Alexandra, always.

1

INTRODUCTION

In the time-worn traditions of the transportation industry, shippers and carriers regard one another as enemies.

There is, to be sure, a certain degree of validity to such a viewpoint. An element of conflict will naturally be present in any vendor-purchaser relationship. The two, after all, are seeking distinctly different things from that relationship; and to a considerable extent each party's success in realizing its own goals must be achieved at the expense of the other.

At the same time, however, viewing that relationship as strictly zero-sum—a gain by one side always resulting in an equal and offsetting loss by the other—is a gross misconception. It overlooks the key reality that, no matter which party comes closest to its own objectives, the relationship, and the business transactions that flow from it, must ultimately benefit both. Without that level of mutual benefit the relationship will simply not exist (or will soon founder if it somehow *does* come into being); for it is only the self-interest of the two parties that impels them to establish and continue that relationship at all.

In transportation, however, this element of mutuality—the attitude that the parties share a common interest in nurturing their relationship—is often forgotten. Instead of working together as, fundamentally, partners in a business transaction from which both expect to derive gains, they dedicate the bulk of their energies to fighting, rather than cooperating, with one another.

Why is the transportation industry characterized by such exceptionally hostile relations between its vendors and its purchasers? There are two basic reasons, which serve to reinforce one another.

First, transportation shares an attribute common to many service industries: The purchaser (shipper) does not really want to buy what the supplier (carrier) has to sell. What shippers want is the *result* of the carrier's service—the "time-and-place value" which transportation, by locating the goods in a particular place at a particular time, lends to the goods being hauled.

This strictly result-oriented approach is to be found in many other economic sectors as well. The individual who patronizes a television repairman, a hairdresser, a dry cleaner, for example, does so under a form of duress rather than by choice; he* wants a working TV, an attractive appearance, clean clothes, and knows of no other realistic way of achieving those objectives. But he—like the transportation shipper—has little or no interest in the operational processes by which his objectives are achieved, and often conceives a sense of resentment at having to depend on others to accomplish them.

Throughout most of the modern history of transportation in the United States this resentment has been encouraged and fostered by the strong government regulation that has existed. Indeed, the very thesis of regulation—that governmental controls are necessary to prevent carriers from taking unreasonable advantage of shippers—seems rooted in the notion that carriers are inherently untrustworthy and perhaps even dishonest by nature.†

The economic regulatory controls that have for so long (until recently) been one of transportation's most important features

*The perceptive reader will note that masculine pronouns are employed throughout this book where reference is made to the generic individual. This should not be taken to imply any prejudgment as to the sex of such an individual; the pronouns are merely used in their old-fashioned sense of referring to any person of unspecified gender, in order to avoid the awkward and intrusive "he or she" locution.

†If this seems an overstatement, consider that the reason given for the extraordinary loss-and-damage liability imposed on carriers, by the jurist who originally enunciated this policy, was that it was necessary "for the safety of all

have themselves also contributed mightily to the negative side of shipper-carrier relations. The basic thrust of regulation was that shippers and carriers were not to be permitted to establish the parameters of their own relationships as they saw fit; rather, government, not the marketplace, would be the final arbiter of "what's best"—and it was the ephemeral "national interest," rather than the interests of the private parties directly involved in any particular relationship or transaction, that would be the controlling factor.

Indeed, so far did this attitude extend that shippers and carriers were by and large prohibited from making *any* significant changes in the way they related to one another without first subjecting their plans to governmental review—a review conducted, moreover, under the presumptive adversary conditions of a legal dispute. In such a milieu it is scant wonder that shipper-carrier relations have been traditionally miserable. Indeed, it would be astonishing if things were otherwise.

Deregulation, however, has put a new face on the situation. The fairly minimal regulatory controls that remain in place impose few limitations on the potential of shippers and carriers to negotiate agreements and "deals" in the marketplace—and offer few protections to those who neglect this aspect of their relationship or handle it poorly. As a result, managers on both sides are having to take an entirely new approach to the shipper-carrier relationship, renouncing the legacy of a regulated past and concentrating on the exigencies of a market-oriented present.

As might be expected where an entire generation of managers is asked to change, virtually overnight, its way of thinking,

persons, the necessity of whose affairs oblige them to trust these sorts of persons *i.e.,* carriers." (*Coggs v. Bernard,* 2 Ld.Raym. 909) The wonderfully disdainful reference to "these sorts of persons" speaks more clearly than any epithets of the low esteem in which the British Lord Holt, who wrote the opinion, held carriers. Admittedly this opinion is now nearly three centuries old; yet the fact that transportation carriers still are burdened with a legal loss-and-damage liability far beyond that imposed on any other businesses, just as Lord Holt prescribed, is a fairly strong indicator that their integrity continues even today, at least in some quarters, to be viewed as suspect.

the results of the first few years of deregulation have been quite mixed. In some relationships, especially those involving "market dominant" rail service (where competitive alternatives are limited or non-existent), carriers have emerged as the clearly dominant partner. In others, mostly where large-volume shippers are taking advantage of competitive overcapacity, the shippers have had the upper hand. And in both types of situations there has been a tendency for the "winners" to exploit their advantage heavily at the expense of the "losers," each clinging firmly to, and even exacerbating, its traditional antipathy toward the other.

In the long term, however, such attitudes are almost certain to prove counter-productive. They neglect the fundamental reality of any unregulated marketplace that a successful business relationship depends on the willingness and desire of both parties to do business together. Even the most captive of shippers can and will, given time and the incentive of an exploitative carrier, develop alternative means of reaching its markets; and even the largest shipper cannot count on an unending overcapacity to serve up carrier after carrier it can bludgeon into submission. Both shippers and carriers who bully rather than bargain must inevitably expect a day of reckoning, when not only do their tactics stop working but their past reputations (as well as their own now-ingrained managerial predilections) make it difficult for them to effectively regroup.

The challenge presented by deregulation, to both shippers and carriers, is thus to shift the basis of their thinking. In a deregulated environment, the successful shipper-carrier relationship will be one in which the parties consider themselves, at least on one plane, not as adversaries bound together only by circumstance, but rather as coequal partners in a business endeavor from which both expect to benefit.

This is, after all, the reality. In the *Rubaiyat of Omar Khayyam* the poet wonders "what the vintners buy / One half so precious as the stuff they sell." And in the final analysis that must be the basis for any business transaction—the buyer placing a higher value on his purchase than does the seller (since otherwise the seller would not sell). Both parties thus stand to gain from their dealings, by receiving from the other something they value more than what is given in exchange.

Once this central truth is perceived, buyer-seller antagonism dwindles to manageable proportions, in today's market-oriented transportation industry as much as in any other economic sector. The relationship now devolves on the question of relative valuation—reaching a middle ground on how the respective parties value what they propose to exchange with one another (here, transportation service on one side and money on the other)—with each seeking a net benefit.

It is through the negotiations process, in one form or another, that this question is resolved.

2

THE VARIEGATED, EVER-CHANGING TRANSPORTATION MARKETPLACE

Most established major industries display an almost rock-like stability in their respective marketplaces.

Year after year more or less than same vendors, occupying more or less the same market niches, are to be found on industry rolls. A relative handful of giants—the Big Three of the auto industry, oil's Seven Sisters, etc.—dominate the market. New entrants are rare, an existing company's departure even rarer. Even the pecking order seldom changes much from one year to the next. And in the few-and-far-between instances when significant change does occur (as through mergers, bankruptcies, substantial shifts in market share, etc.), it sends waves of anxious tremors throughout the industry.

But though it certainly qualifies as an "established major industry" by any measure, transportation exhibits little of this stability. Some firms are, of course, bigger than others; but there's no clear dominant group in transportation's marketplace. Companies come and go in the industry, and move up and down the economic ladder, in vast numbers. Change is neither surprising nor seriously troubling; it's the norm.

Regulation, of course, is the reason—decades of government regulation that caught transportation in its infancy and held it there in suspended animation. Quite simply, transportation was not allowed to progress through the normal stages of development that have led, in other economic sectors, to a naturally stabilized marketplace.

A new industry is born out of some sort of shift in the existing order—a technological advance, a social realignment, a change in cultural values, etc. Typically it enjoys an early boom, which attracts new entrants as a flame does moths. Overpopulation sets the process of economic Darwinism into motion, producing a succession of "shake-outs" that winnow out all but the fittest to survive. Gradually the boom-and-bust cycling levels out, leaving a stabilized core with a solid grasp on the market.

Before that happens, though, the sequence of expansions and forced contractions—a steady flow of entries into and exits from the market—will inevitably have a highly disruptive impact on the industry's consumer population. For various reasons in transportation (they differed from mode to mode), it was deemed socially and politically unacceptable for the consumers of transportation services to endure such disruptions. So the government stepped in to impose its own artificial stability through regulatory controls and, in so doing, abruptly arrested the industry's normal course of development.

Volumes could be, and have been, written on the genesis and growth of the deregulatory movement of the 1970's. It's enough, here, merely to say that large-scale deregulation did take place, with most or all of the constraints on the transportation marketplace being removed.

The effect was to release the industry from its state of suspended animation at more or less the same point in its development at which it entered that state—in other words, still in the earlier phases of its expansion/shake-out cycling. And so we are treated to the paradox of a mature industry that is nevertheless as volatile and unstable as a new one—much as if a middle-aged man suddenly began cavorting like a 10-year-old.

Because of inherent differences among them and their markets, the various transportation modes have reacted differently to their new freedom; and different responses are called for from those participating in the marketplace, whether as transportation suppliers or consumers. In each case, however, the industry's newfound marketplace volatility affords both problems and opportunities for carriers and shippers alike.

Railroads

In a very real sense the railroads are the Rip van Winkle of the transportation industry.

Oldest among the modern transportation modes (the first railroad was built over 150 years ago), they likewise have the longest history of regulation. The first state "Granger Laws" were enacted over a century ago, and controls at the Federal level date back to 1887. For nearly a hundred years railroad management was forcibly anesthetized by government policies that constrained its power to make and carry out business decisions. Not until 1980, with enactment of the Staggers Act, were the stultifying government restraints at last (largely) lifted.

Like Rip, the railroads have awakened to a world vastly different from the one they knew before. But unlike Rip, they have no option to ignore most of the differences and sit around gossiping with old cronies and playing with the new generation of children; they must adjust to the new order and compete in the new marketplace.

A substantial part of that adjustment involves streamlining their "plant," ridding themselves of unprofitable branch lines and services that for decades have overloaded their cost structures. Regulatory policies of the past were ultra-conservative on this score, often obliging the carriers to continue money-losing operations as a matter of "public convenience and necessity." Under the Staggers Act, however, no railroad may be prohibited from abandoning any line or discontinuing any service that isn't turning a profit.

Indeed, current Interstate Commerce Commission* policy

*As of when this was written, there was pending before Congress a proposal to abolish the ICC altogether. This proposal was tied closely to total economic deregulation of other transportation modes subject to the Commission's jurisdiction (as is discussed below). But it was equally clear that there would be no substantive change in rail regulation under the Staggers Act; rather, in the event of the ICC's demise these regulatory responsibilities would merely be transferred, largely intact, to other Federal agencies. In the event this proposal has been adopted by the time the reader sees these pages, he

(which has been ratified by the court system) even permits railroads to do away with money-making lines if the profit margins aren't high enough. The ICC allows carriers to include, in costing out undesired branch lines, "opportunity costs"—the revenues they are foregoing because capital tied up in a line isn't available for alternative investment, based on a fixed return-on-investment margin set annually (based on current economic conditions) by the Commission. If the line's margin above other, actual costs is less than this figure, it officially qualities as "unprofitable" and may be abandoned.

Prior to 1980, shippers were encouraged to file formal "protests" with the ICC any time a carrier proposed abandonment of a line or discontinuance of a service in which they had an interest. They did so liberally, for reasons both important and frivolous (in one actual case a farmer opposed halting a passenger train he never rode simply because he "liked to hear the whistle blow" as the train steamed past his fields); and in many cases they succeeded in blocking the action. The ICC must still approve abandonments and discontinuances, and shippers may still file protests; but unless the line or service can be proved to be earning an adequate profit, the Commission has no legal authority to require its continuance, and the protests will prove futile.

The present-day shipper's recourse to prevent abandonments/discontinuances that could harm his interests lies in the marketplace, not regulation. The Staggers Act affords him three alternative means of averting that harm:

(1) He may offer to subsidize continued operation of the line or service, making up, in periodic payments to the carrier, any revenue shortfall. Provided the subsidy is sufficient to let the carrier break even (taking into account, of course, opportunity costs), the law does not permit the carrier to turn down the offer. To protect the shipper from unreasonably high subsidy demands,

---

should thus mentally substitute the name of the agency(ies) to whom the regulatory authority has been assigned wherever the Commission's present and/or future role is mentioned; otherwise, matters will remain unchanged.

the law further empowers the ICC to fix the amount of the subsidy if the parties can't reach a negotiated agreement.

If the operation involved is a service—such as a switching activity, a scheduled passenger train, etc.—this is the only alternative to discontinuance. If it is a branch line that the carrier proposes to abandon, however, two other options are available:

(2) The protesting party may buy the line and operate it himself. Again (subject to certain restrictions as to timing), the law does not allow the carrier to turn down such a purchase offer. But it does insist on a fair price—either (a) the "going concern" value of the line (*i.e.*, its value as an operating rail line), or (b) its net salvage value, whichever is higher—and lets either party appeal to the ICC to set the purchase price if they can't agree privately. The selling railroad is also obliged to provide that now-private line with adequate connections, switching service, etc.

(3) He may locate some third party who, on an entrepreneurial basis, will purchase and operate the line as an independent for-hire carrier enterprise. Minimum terms of sale are as previously described; the ICC, again, serves as the arbitrator of last resort; and appropriate connections must be afforded by the selling railroad.

In the first few years following enactment of the Staggers Act this third alternative accounted for a considerable increase in the number of small rail carriers. In some cases the new operators are primarily hobbyists, railroad buffs fulfilling a lifelong ambition to run their own lines with only a secondary concern for profits. But in many others they are hard-headed businessmen who expect to make money from their ventures; and often they succeed. In fact, several companies have sprung up that make a practice of acquiring and operating small lines under these circumstances; and others, although not themselves involved in rail operations, broker such deals with interested parties and offer financial investment and/or managerial and other services to the new carriers.

How is it possible for a newly minted railroad, with probably inexperienced management, to earn a profit from a line some much larger, expertly managed carrier has given up as a money-loser? To be sure, the new short-line road may charge somewhat higher rates; but if it were only a matter of that, the larger carrier could readily have increased its own rates. The key lies instead in some significant cost reductions—some "paper," some in real dollars—the small carrier can realize.

• Standard costing ascribes a *pro rata* share of a carrier's overhead expenses to each individual line or operation. In many cases the new, small carrier, with only a single short line to run, can pare such costs down to virtually nothing. More than one such railroad exists whose president is also chief locomotive engineer, whose office is the basement of someone's home, whose clerical personnel are the owner's teen-age children, etc. Even where such carriers are operated in a less homespun manner, overhead may still be appreciably lower than the previous major-railroad owner incurred.

(Of course, the big railroad hasn't actually eliminated much, if any, of the overhead expense it previously charged off to the line in question. Incremental cost savings that can be realized by abandoning a line, or even many lines, will always fall far short of the overhead costs ascribed to the line(s), for a variety of reasons. Instead, what has actually happened is that most of the overhead has simply had to be reallocated among the carrier's remaining lines and services, adding to their cost burden and perhaps pushing some of them over the brink into nominal unprofitability.)

• New short-line carriers are not required to adopt the union contracts under which the line's previous owner was obliged to operate. Since labor alone accounts for just a shade under 50% of total rail operating costs, the opportunity to avoid union wage scales, restrictive work rules, etc., can allow considerable real-dollar savings.

• The new carrier's management is accorded a completely and permanently unregulated status. Although the post-Staggers Act regulatory environment imposes relatively few obligations and burdens on rail carriers, regulatory requirements still account for some carrier costs—and also saddle carriers with certain duties and liabilities *vis-à-vis* shippers.

• Finally, the shipper or shippers served by the line is (are) apt to be more cooperative with the new small carrier than with its major-railroad predecessor. It or they may be more willing to accept rate increases without diverting traffic, more amenable to compromise on service needs, less strong-willed about loss-and-damage claims, etc. In part this is straightforward self-interest; any realistic shipper knows the short-line carrier represents its last hope of retaining rail service (short of, of course, operating the line itself), and in some instances shippers are also major investors in the new carrier entity. In part, too, it may be a matter of improved customer relations by the new carrier (which has only one or a few shippers to please, rather than countless thousands), and/or the natural human sympathy shipper managers may feel for small-business suppliers.

Whatever the reasons, a surprising number of these new short-line railroads appear, at least from early indications, to be economically viable entities. Considering that each began with a line or lines jettisoned by larger carriers as both unprofitable and lacking future profit potential, this is well worth remarking, and indicates that this option is worthy of a shipper's consideration where it may face abandonment of a line it wants to keep in service.

Before leaving the subject of railroad abandonments, one additional protection incorporated in the Staggers Act warrants mention. From time to time it has happened that a major railroad, for various managerial reasons of its own, wanted to abandon a line or discontinue a service that obstinately persisted in generating too much revenue to meet regulatory criteria. In such cases it has not been unknown for the railroad to purposefully allow the line to deteriorate or downgrade the service, in the hopes of driving away enough traffic (and hence revenues) to qualify it, in regulatory terms, for abandonment/discontinuance.

The law provides in such cases for shippers to ask the ICC to review the situation and (provided the results of the investigation bear out the shipper's contentions) order the railroad to improve things. Continued failure by the railroad to do so subjects the line to potential forced sale to either the shipper itself or a third party, as described above.

In the event, this latter statutory provision has proved largely unnecessary. In the past more-regulated era, when carriers could be and not infrequently were required to continue marginally profitable lines or services, or even unprofitable ones where the losses weren't deemed excessive, carriers did have an incentive to take such steps. Now that the law guarantees them a substantial profit from any line or service (or else frees them from the obligation to continue operating it), however, the tactic of intentional downgrading is both unnecessary and, from a business point of view, unsound.

Just as regulation encumbered the railroads with an outsize, economically burdensome operating "plant," so it also imposed on them a rigidly structured set of formulae for sharing revenues on interline movements which many felt was egregiously unfair.

In the nature of things, the majority of all U.S. rail service involves interline movements; given any pair of points at random, the probability is that the same carrier will not serve both so that, if goods are to move between them by rail, at least two and possibly more carriers must participate in the movement. Prior to 1980 the Interstate Commerce Act effectively obliged carriers to agree amongst themselves on "joint rates and through routes," so that the shipper would be assessed a single-factor rate, roughly comparable to what he might pay if single-line service were available, for such movements.

Not surprisingly, railroads constantly bickered with one another as to how they would share out the revenues from such joint-line traffic. Because of this, during the 1930's the ICC conducted an extensive investigation and set out specific formulae by which (if they could not agree otherwise) these revenues would be divided.

It should go without saying that over ensuing decades these divisions formulae became outmoded. Railroad traffic patterns, cost factors, etc., changed, often radically; but the formulae didn't. Efforts to recast the formulae in line with more up-to-date conditions were mounted in the 1950's and '60's, but were aborted short of completion for various reasons. By the time the Staggers Act was being considered in Congress, many carriers (especially in the eastern part of the country) could accurately complain they were losing money, often a lot of it, on the interline traffic they handled.

Congress thus incorporated in the act a provision giving them an alternative. Any carrier that wasn't receiving at least a 10% margin over its direct costs from a particular interline routing was allowed to cancel its participation in joint rates over that route. Although provision was made for shippers and even connecting carriers to protest such cancellations, the ICC has rarely overruled the carriers' decisions in this regard; and a substantial portion of joint rates (especially, again, involving eastern carriers) were accordingly cancelled during the first few years of the Staggers Act's effectiveness.

Cancellation of these joint rates resulted, of course, in a considerable increase in the charges shippers must pay. With no joint rates, charges are based on a "combination of locals"—that is, the sum of each participating carrier's rate between the point at which that carrier receives the freight (whether at origin or from an interline carrier) and the point at which its movement terminates (at destination or an interline point). Since these individual-carrier rates are based on the premise that each carrier is performing a full pickup and a full delivery service—and since, moreover, they cover much shorter distances than the full interline route, thus reducing the effect of the normal distance "taper" on rate scales—they often add up to a lot more money.

Some carriers, too, have been hurt by this trend toward cancellation of joint rates. In particular this question seriously concerns short-line railroads whose business, obviously, is virtually 100% interline in nature. And larger carriers, too, have seen traffic diverted to competing rail routes and even other modes, sometimes all but overnight, when joint-rate cancellations pushed rate levels uncompetitively high.

In some instances, carriers have expressed a willingness to publish "proportional" rates which, although covering only a particular carrier's leg of an interline movement, would be pegged lower than true single-line rates when that carrier was merely originating or terminating an interline movement (and perhaps lower yet when it served as a "bridge" carrier in a movement involving three or more railroads). Others, although unwilling to publish joint rates in their common-carrier tariffs, have established them *ad hoc* as part of contractual arrangements (*see Chapter 8).*

And in some instances another marketplace trend in the railroad industry—mergers and consolidations—has made the question moot by combining the prior partners to an interline movement into a single corporate entity.

It has long been apparent that the industry's alignment into 35-plus basically regional carriers was badly out of step with the times. Such an alignment made good sense perhaps up to World War I or so; but since then the steady growth of longer-haul transportation demand, and the inroads in shorter-distance, regional traffic by motor carriers, have made it increasingly obvious that the railroads could realize significant economies of scale through well-conceived unifications.

Perhaps the most grandiose such proposal came, in the early 1970's, from an ICC staff official, one Nathan Klitenic. Assigned to preside over a very complex merger case involve competing proposals absorb the Chicago, Rock Island & Pacific Railroad, Klitenic went far beyond his brief in a voluminous report, suggesting that the numerous major carriers then providing service west of the Mississippi be consolidated into, essentially, just three major systems.

Of course, Klitenic didn't trouble to check his proposal out with the executives of the roads involved, most of whom greeted it frostily. But the biggest impediment to implementing it or any other major rail consolidations—as it had been throughout almost the entire regulatory era—was the requirement for prior ICC approval.

To begin with, approval wasn't ever that easy to get for anything but relatively inconsequential rail consolidations. The law obliged the carriers to prove the transaction would enhance the ubiquitous "public convenience and necessity," and the Commission, throughout most of its history, took this commandment (inspecific though it was) seriously. Both shippers and competing carriers were afforded an opportunity, even encouraged, to oppose mergers; and the ICC took them seriously, too. And even where it did allow a consolidation to take place, the agency often imposed so many conditions—to protect labor against job losses (even through attrition in some instances), to protect competing and/or interlining carriers against traffic diversion, to ensure continued service on lightly used branch lines, etc.—as to disallow many or most of the potential economies of the transaction.

The most imposing obstacle of all, though, was not any action the ICC took, but rather the length of time the carriers had to wait for it to act at all. Delays of three, four and five years were commonplace; in the Rock Island case over which Klitenic presided (admittedly an unusually difficult one), the decision took 11 years to come, by the end of which time the Rock Island itself was in bankruptcy and the original proposals so out-of-date as to be unusable (in fact, none was ever consummated).

Indeed, delays of much shorter duration than that can have the same impact on proposals for virtually any form of corporate consolidation. Negotiations leading toward a merger, acquisition, etc., are always exceedingly delicate, involving a great many different individuals and groups, *all* of which must come to full agreement before the deal can be closed. That agreement, naturally, will be based on conditions as of a given point in time—asset values, liabilities, operating and marketing forecasts, stock prices, etc., etc. The longer beyond the time of that agreement it takes to consummate the transaction, the more likely it is that key features will change, to the point that one or more of the key parties will withdraw his (or her or its) agreement. When you combine this with the fact that delay in consummating an announced consolidation also gives opponents to that consolidation (and there almost invariably will be opponents) a chance to muster their forces, the probability that the transaction will actually take place diminishes almost exponentially with the amount of time required to complete it.

As a result, railroad consolidations prior to the mid-1960's for the most part took place only in dire circumstances—when, for example, the alternative would be insolvency and likely dissolution of one of the carriers involved. Then came the Penn-Central merger and the collapse of the surviving carrier alarmingly soon after the unification, and the regulatory view of consolidations grew even (if possible) dimmer.

Although keeping in place the requirement for prior ICC approval, the Staggers Act (and, to a lesser degree, the Railroad Revitalization and Regulatory Reform ("Four-R") Act four years earlier) changed all this dramatically. First, it reversed the prior legal/regulatory presumption about mergers; now any consolidation agreement is presumed to be consistent with the public interest unless proved otherwise. Second, it stripped away most of the Commission's authority to impose protective conditions that might impede the surviving (consolidated) railroad's chances of realizing key economies and efficiencies. And third, and perhaps most important, it limited to an absolute *maximum* of 30 months the time the ICC might take to consider a consolidation proposal, and required that most be decided in much less time than this.

The result was a virtual epidemic of rail consolidations in the years immediately following enactment of Staggers. Indeed, the merger movement tended to feed on itself; because of the lack of ICC-imposed protective conditions, the consolidation of two carriers frequently deprived a third of considerable interline traffic (and revenues), and hence forced it to scurry around hastily for its own merger partner in domino-like sequence.

For antitrust reasons, most railroad mergers have been of the end-to-end variety; that is, the merging carriers served different regions and interlined traffic with one another over one or a few interchange junctions. (The Justice Department, which enforces the antitrust statutes, tends to oppose mergers of direct competitors—*i.e.,* carriers with substantial overlapping trackage and service—as being anticompetitive.) The result has been creation of a new industry structure in which much longer single-line hauls, and much less interlining of traffic, are becoming the norm. As these words are written, the end of this trend appears nowhere in sight; indeed, the likelihood is that there will be one or more truly transcontinental railroads, capable of coast-to-coast single-line service, within fairly short order.

The final important effect of deregulation on the shape of the railroad marketplace has been the explosive growth of rail contract service.

Prior to 1978 there was no such thing as contract rail carriage; all rail service was by definition common carriage, governed by a host of regulatory requirements and legal constraints. In that year, however, the ICC abruptly reversed 90 years of prior policy decisions on this question, in a ruling subsequently affirmed by Congress in the Staggers Act; and by the end of 1985 contract service was had reached a level estimated at 40% or more of all U.S. rail transportation, with no signs of abatement in the rate of growth.

Any railroad, and any shipper, are eligible to enter into contractual agreements; the only limiting factor is their mutual willingness to do so. Such contracts entirely override regulatory and (in the main part) common law, and may include virtually any provisions to which the parties agree.

Most commonly, rail contracts are between one carrier and one shipper and involve single-line service. Interline contract service, however—involving through service by two or more carriers—is also clearly permitted by the law, and many such contracts have been negotiated.

It is too early (as of the time this is written) to assess a final trend in the railroad sector which began to emerge only several years after the Staggers Rail Act became law: expansion into other modes of transportation. Freed from decades-old restrictions that limited them to steel-wheel-on-steel-rail operations and prohibited them from engaging (either themselves or through affiliates) in other types of service, several railroads began acquiring established motor carriers (and one a barge line). The extent to which this might affect their operations long-term was uncertain as of the time this was written (*see the discussion of intermodal transportation below*).

Motor Carriers

As of when this was written, most for-hire trucking service remained nominally subject to government economic regulation.

In a real-world sense, however, regulation has become such a toothless entity at the federal level (and in many states as well) that it may be regarded as *de facto* nonexistent in any substantive way—more a matter of red tape than of meaningful requirements and restrictions.*

Thus, for all intents and purposes it is the marketplace, and not government that dictates the relations between motor carriers and their shippers.

Technically, motor carriers require regulatory permission before they may enter that marketplace.† The standards by which regulatory law is administered, however, have been dramatically weakened in the 1980's. At the Federal level compliance is, again, a matter more of form than of substance; it is rare indeed that any carrier's (or would-be carrier's) request for authority to provide for-hire trucking service is turned down or even reduced in scope by the ICC. With regard to intrastate service, standards vary considerably from one state to another, but in general are much less stringent than in the past.

Because of a peculiarity of the deregulatory statutes, the distinction between inter- and intra-state operations is important in trucking whereas it is insignificant in rail transportation. The Staggers Rail Act obliges all states to bring their regulation of intrastate rail service exactly in line with Federal standards, and deprives any state failing to comply of its regulatory authority (which in such circumstances is subsumed by the ICC).‡ The

*Indeed, the probabilities appeared strong, when this was written, that even this *pro forma* residue of regulatory controls would soon be eliminated at the federal level, a step some states had already taken.

†To avoid meaningless repetition, the phrase "as of the time this was written" will not be repeated in this discussion. All of the following comments on the details of present-day motor carrier regulation should, however, be viewed by the reader as if modified by that phrase.

‡As of this writing, only one state—Texas—had been so recalcitrant about adopting Federal regulatory standards that it had lost its authority to regulate intrastate rail service. Other states, however, remain subject to this legislative mandate, and stand to likewise lose their regulatory authority should they at some future time deviate from Federal standards. Likewise, the Texas Railroad Commission could regain its authority if state law and rules were at a future time changed to comport to those standards.

companion Motor Carrier Act of 1980, however, contains no such proviso, leaving the individual states free to legislate and implement regulation as they may choose.

The result is that, although rail regulation has achieved almost total uniformity throughout the country, motor carrier regulation is very *non*-uniform. At one extreme several states have totally deregulated trucking service, allowing carriers to operate in the intrastate marketplace without any regulatory oversight at all. At the other end of the spectrum, a few states maintain such tight controls that it is extremely difficult for a new carrier to enter the market at all or for an existing one to expand its market coverage.

These various state standards apply, however, only on traffic that remains within a single state's borders for the entirety of its origin-to-destination movement, including connecting moves. If a shipment, for example, originates at a point in one state, moves to a point in another state, and is then interlined for further movement elsewhere in the second state, the *entire* through movement—including the second portion, which itself never crosses state lines—is deemed to be in interstate commerce. Likewise, traffic that both originates and terminates at points in the same state, but traverses a route that passes through another state (provided there are valid operating reasons for using that route), is considered interstate in nature. And all interstate movements are, of course, subject to Federal, not state, regulatory standards.

Federal law has always exempted certain types of truck transportation from any form of economic regulation—exemptions that are basically unchanged from the pre-1980 era, although the Motor Carrier Act of that year did enlarge them in some respects. The chief exemptions are these:

- Transportation of various specific types of goods: Agricultural commodities, including seeds, plants, and livestock and poultry feed; fish and shellfish products and by-products; used empty shipping containers; wood chips; and decorative stone.

- Transportation that is "incidental" to air freight movements—pickup and delivery service, and occasional substitute motor-for-air movements when, for example, adverse weather conditions make air service impossible.

• Supplementary for-hire service by agricultural cooperatives—service they offer to the public at large to improve the economy and/or efficiency of their primary operations on behalf of their farmer members (subject to certain restrictions as to the volume of non-farm-related traffic they may haul for nonmembers).

• Private carriage *(see below)*.

As to all of this service, there are no regulatory controls whatever at the Federal level; market entry is strictly a managerial decision on the part of the carrier.

With regard to all other interstate truck transportation, carriers wishing to enter the market must first secure, from the ICC, either a "certificate of public convenience and necessity" (as to common carriage) or a "permit" (as to contract carriage). These certificates and permits identify, albeit in very broad terms, the types of goods their possessors may haul and, as to common-carrier certificates, the points and/or areas they may serve (contract-carrier permits are not subject to geographic restrictions).

When these requirements were originally put in place, in 1935, the trucking industry was in competitive chaos. Still in its fledgling stages of development, it was characterized by a large number of small companies battling for traffic that, in the depths of the Depression, simply wasn't adequate to support anything like the capacity that then existed. The industry was ravaged by rate warfare that imperiled its financial stability, while shippers had to struggle with unreliable and often unsatisfactory levels of service.

The industry's problem wasn't hard to diagnose: There was simply too much competition for too little traffic. In another industry the situation might have been left to resolve itself through the normal process of economic attrition, with the unsuccessful competitors dropping out and thereby helping stabilize the marketplace for the survivors. But there are few market barriers to entry into the trucking industry (low capital investment, no requirement for major facilities, etc.), and attrition wasn't producing results very quickly; as fast as one trucker failed, another—or even two or three—would arise to take its place.

Because of the importance of transportation in the economic infrastructure, Congress decided there was a need to hasten the stabilization of the trucking industry. It thus interposed barriers to entry into the industry, in the form of the requirements for certificates and permits before a company might provide for-hire motor carriage; and it instructed the ICC to enforce those requirements stringently in order to curtail the excessive market competition.

This philosophy was followed for over 40 years. The regulatory presumption was that there was plenty of existing trucking service in any given market segment; and a would-be entrant was obliged to affirmatively disprove that presumption before the ICC would grant it authority to become a new competitor in the trucking marketplace. Gaining a common-carrier certificate was an extremely difficult task in most cases, while even contract-carrier permits (which at that time were far more restrictive in nature than is now the case) were not easy to come by.

By virtue of a key statutory loophole, the 1935 statute did leave some room for further market development of the trucking industry. Any firm that could prove it was, at the time the law was enacted, already serving a particular market sector was automatically entitled to "grandfather" authority—certificates or permits—to continue doing so; it was up to natural economic forces to choose among competitors in the already oversaturated marketplace of that time. As a result, the trucking industry did make some further progress—albeit, in light of the limits imposed on new competition, very lethargic progress—toward coalescence and consolidation.

Even so, protected as they were from new competitors, a vast number of truckers were able to maintain at least a foothold in the industry—far more than would have been the case had the industry been allowed to develop unencumbered by regulatory constrictions. By 1980 there were still an estimated 16,000-17,000 for-hire motor carriers competing for regulated traffic, and even the largest of them commanded but a minuscule share of the total market.

The Motor Carrier Act of 1980 wrought vast changes in the regulatory scheme even while nominally leaving in place the requirement that new market entrants first obtain certificates or permits from the ICC. Quite simply, it reversed the underlying presumption of the previous law that sufficient competition already existed in any market segment. Under the new statute, the presumption was that the marketplace *did* need new competitors—as many as wanted to enter it—unless proof to the contrary could be adduced; and then the requirements for such proof were made, through regulatory interpretations and policies, so stringent that it is, as a practical matter, impossible to meet them.

Thus, carriers already in a market overnight lost their regulatory protection. In short, the door to the motor transportation market, barely ajar prior to 1980*, abruptly swung wide open. Any carrier, simply by meeting a few mostly formalistic requirements, could secure the requisite certificate or permit to enter the market. Predictably, there was a rash of new entrants, to the extent that the number of carriers authorized to haul regulated traffic approximately doubled in the five years leading up to the end of 1985.

For the first two to three years after enactment of the 1980 statute many established truckers failed to recognize how radically matters had changed. As they had done prior to 1980, they continued to file "protests" against the applications of new entrants who proposed to compete with them, arguing that they (the protesting carriers) were meeting all the proven needs of the marketplace already. They contended that additional com-

---

*In actuality, the transition formalized by the Motor Carrier Act of 1980 had begun several years earlier as a result of sharp changes in ICC policy. During the late 1970's the Commission had greatly liberalized its certificate/permit processes to such a degree that motor carrier entry barriers had already been vastly reduced even prior to enactment of the law. This had led, however, to much controversy and a considerable amount of litigation, which the new statute did much to resolve.

petition would only serve to dilute available traffic and revenues, to the detriment of all in the market.

Such protests, however, were invariably unsuccessful. Provided only that the applicant carrier had properly completed the paperwork, it was, as a practical matter, guaranteed success in its quest for a certificate or permit to provide the service it was proposing. By the end of 1985 the message had been communicated to even the dimmest of carriers, and virtually all applications for certificates and permits were being routinely processed by the ICC without opposition.

The competitive pressures resulting from this sudden relaxation of long-entrenched entry restrictions have had profound effects on the motor carrier marketplace. The most evident, and most spectacular, has been a procession of bankruptcies involving some of the largest carriers in the industry.

To a considerable extent these bankruptcies resulted from a confluence of the deepest economic recession since the Great Depression of the 1930's with the advent of motor carrier deregulation. Carriers that might well have been able to survive the exigencies of deregulation alone found themselves powerless to withstand the deregulation-induced pressures in concert with the curtailment of traffic caused by the 1981-83 recession. And even after the nation as a whole was embarked on a powerful economic resurgence (beginning about the middle of 1983), many major carriers had been so severely weakened by the recession that they ultimately succumbed.

Most vulnerable were the carriers with strong Teamsters Union representation. Over the decades of regulation the Teamsters had taken advantage of the protected-market status of carriers to push wage and benefit levels far above those of comparable, non-regulated industries. The carrier/employers, without severe marketplace pressures to maintain stringent caps on their costs, had been easy prey to the tough negotiating tactics adopted by the Teamsters; given the competitive controls of the era, the carriers could simply pass increased labor costs through to their shippers without competitive harm. Wage/benefit levels in the motor carrier industry had risen to the second-highest of any major U.S. industry, exceeded only by the airline sector (*see below*).

When deregulation abruptly opened up the market to largely unrestrained competition, neither the carriers nor the union recognized how much their past practices had crippled their market-competitive potential. Motor carriage has always been an extraordinarily labor-intensive industry; labor costs amount to as much as 70% of the total operating costs of any given motor carrier. Thus, new non-union entrants—or existing non-union firms with expansionist ideas—were able to exert extraordinary competitive pressure by simply reducing labor costs to levels well below union scale (though still comparatively high in terms of the general labor scale outside transportation).

Union-dominated carriers accordingly began dropping like proverbial flies. Some—generally the largest, and with the greatest financial resources (to enable them to withstand the calamity of the 1981-83 recession)—were able to surmount the pressures and, by achieving economies in other operating areas, even to grow and prosper. Others, although trembling on the brink of insolvency, managed to remain at least afloat—often bolstered by a curious legal situation which, by according union pension funds first (and heavy) call on their assets if they declared bankruptcy, discouraged their creditors from pressing them excessively. But a significant number tumbled into bankruptcy; and the prospect (again, as of this writing) of a relaxation in the pension-fund law made it evident that more would be following suit over the years.

In addition, beginning in late 1984 motor carriers found themselves under still further pressure from another, unexpected source—their insurance underwriters. This problem stemmed from a provision of the 1980 deregulatory statute that sharply raised commercial truckers' requirements for liability coverage. Concerned about the impact of more relaxed motor carrier entry policies on highway safety, Congress had sought to make the safety situation self-policing; it reasoned that, with much larger sums at risk, insurors would either withdraw coverage from unsafe operators or raise their premiums to astronomic levels, either way quickly driving these operators off the roads and out of the market.

Logical though that approach might be, it failed to take into account the fundamental irrationality of the competitive marketplace. Just as it did in the motor carrier market itself, the deregulation-induced industry expansion created an atmosphere of intense competitiveness in the insurance sector as well. For the first few post-1980 years insurance was easy to get for virtually any existing or would-be trucker, and premium costs were at bargain-basement levels.

Inevitably, the day of reckoning came. More than a few insurance companies were forced, by simple economics, to either sharply curtail or actually cease writing motor carrier liability policies; some even suffered complete economic failure. Abruptly the buyer's market for motor carrier insurance became a seller's market, with premiums commonly doubling or tripling literally overnight and coverage hard to come by even at these prices. For a good many trucking concerns this was the final nail in their coffins; already beset by dwindling profits as a result of their own competitive situations, they no longer had any capacity to absorb cost increases of the magnitude their insurors were presenting to them, and so withdrew from the market—either voluntarily or via the bankruptcy route.

As of this writing the situation remains far from stabilized. Insurance pressures on the carriers continued into the early part of 1986; the pension-fund prop was weakening; and the trucking market remained severely burdened by serious overcapacity. It is clear that the motor carrier industry still has a long distance to travel before reaching anything approximating competitive equilibrium—and that the way will be speckled by further business failures on a substantial scale.

Meantime, the merger/consolidation movement that has characterized the post-deregulation rail sector never really materialized in motor carriage. A carrier encountering economic difficulties had little to offer prospective merger partners (unlike the railroads with their proprietary trackage bases), and thus would generally be allowed to fall into insolvency rather than being accepted by a more financially stable firm as a merger partner.

The departure of these bankrupt carriers is leaving, obviously, substantial "holes" in the trucking market, into which, in large part, the larger and better-financed survivors naturally flow. There is, thus, a tendency for the larger companies to considerably enlarge their market share as these bankruptcies open up competitive opportunities.

At the same time, there has been a proliferation of smaller carriers whom regulatory law no longer bars from entry into the marketplace. Notwithstanding the numerous bankruptcies, the roster of "authorized" (by the ICC and/or state regulatory bodies) motor carriers about doubled over the first five years of deregulation. The vast majority of these new entrants are extremely small in size (the "mom-and-pop" variety); but a few have been able, by one means or another, to attain considerable stature in the newly volatile marketplace.

Thus, the post-deregulation motor carrier market has been characterized by two seemingly contradictory trends—on the one hand a marked consolidation of "market power" into the hands of a relative few major carriers and, on the other, a dramatic increase in the number of competitors in that market. In other words, there are a lot more carriers in the industry; but the bulk of the motor carrier market is becoming concentrated in the hands of a relative few.

Equally paradoxical is the significantly increased homogeneity of the motor carrier industry—a marked tendency of carriers to operate, and to market themselves, as transportation generalists. One might expect that the increasing population of the marketplace would naturally foment increased specialization, as each carrier sought to carve out its particular market niche in order to reduce the competitive pressure on it; but in the early years of deregulation, at least, just the opposite has occurred.

Once again, one must look to the particulars of the regulated past for an explanation. As has already been discussed, market entry was strictly limited during those years, with existing carriers accorded absolute preference so long as their service did not exhibit major deficiencies. This meant that would-be entrants were best advised to seek out obscure market segments where they would not have to overcome the opposition of existing carriers in order to secure the requisite operating authority from regulatory authorities.

Thus, by an unspoken collaborative effort, government and industry contrived to subdivide the marketplace into countless artificially delineated fragments. In terms of both the commodities they might transport and the geographic areas they might serve, the carriers, with governmental regulators egging them on, created "new" market segments based on hairsplitting distinctions. One carrier might haul this product but not that one; another might haul goods if they were bundled onto pallets, but not if they were unpalletized; a shipper might call on one carrier for its outbound movements but have to employ another for inbound; and so forth.

Unsurprisingly, deregulation engendered a pendulum-like reaction to this long-standing enforced specialization. Carriers that had been constrained by such limitations were able to escape from them; every carrier could readily secure authority to haul any goods from and to any points. Wisely, most carriers did not instantly seek to expand their services quite this broadly; even so, however, the average carrier took advantage of the opportunity to broaden its operations considerably, in terms of both commodities hauled and geographic areas served.

Quite obviously, few (if any) carriers can hope to be "all things to all men shippers" over the long run; it is inevitable that specializations must develop. In certain areas this is already in evidence, as where specialized equipment, facilities, etc., may be required to perform particular types of service (liquid and dry bulk cargoes, shipments of extraordinary weight or bulk, parcel and other small-shipment service, etc.). If the motor carrier industry is to follow the pattern of other industries—and there is every reason to expect it to in the normal course of its economic maturation, now that the stultifying restraints of regulation have been lifted—much greater specialization may be expected to emerge, as carriers settle into chosen or *de facto* market niches.

In another area, however—the long-existing dichotomy between common and contract carriage Mthe likelihood is that the deregulation-induced move away from specialization is permanent. During the decades of regulatory control these two forms were kept rigorously separate. Regulators such as the ICC saw their primary objective (correctly, based on clear legislative intent and standards) as being to foster the development and growth of a sound common carrier network to serve the public at large. Accordingly, they sought to discourage alternative forms of transportation, including contract carriage. (This discouragement was even more in evidence when it came to private carriage; *see below.)*

This policy took two forms. The first was that (with only a handful of exceptions) a carrier was permitted to provide only one form of service; it might be either a common carrier or a contract carrier, but "dual operations" as both were prohibited. Secondly, the number of different shippers a contract carrier might serve was arbitrarily limited; in general contract carriers were bound by the ICC-imposed "rule of eight," meaning they might have simultaneous contracts with no more than that many shippers.

This meant a carrier was obliged to choose one form or the other; and if it had significant expansionist ambitions, common (not contract) carriage was the only choice that made sense.

Indeed, toward the end of the regulatory era contract carriage had become for the most part a sham—a mere device to avoid regulatory limitations. If a carrier wished to begin service to a shipper already served by existing carriers, it stood a much better chance of gaining the requisite operating rights if it proposed "specialized" contractual service than if it sought broader common-carrier authority. As a result, the contractual relationship between shipper and carrier often existed "more in the breach than in th' observance"—a formality with little real-world substance behind it.

The Motor Carrier Act of 1980 swept away the restrictions that had led to this situation, with a marketplace impact whose full implications are still uncertain. Within but a few years after its enactment the long-standing line of demarcation between common and contract carriage had almost completely evaporated. Virtually every common carrier of any significance in the marketplace had established a contract-carriage division or affiliate; and except for the very smallest (mostly those with close ties to a single shipper), contract carriers reciprocated by expanding into common service. Contractual relationships became a matter of choice between shippers and carriers, rather than a byproduct of regulatory mandate.

It is difficult to assess how this has affected (or will affect) the relative importance of common and contract carriage in the marketplace. Certainly there has been (at least in the early years of deregulation) no dramatic expansion of contract service, as there was in the railroad sector *(see above);* that is, of course, only to be expected in light of the vastly greater range of choice available to shippers using motor instead of rail service. On the other hand—and this *is* perhaps mildly surprising—there has been no discernible diminution in contract motor service, either; if anything, in fact, the volume of traffic moving contractually has moderately risen.

At the time this was written, however, it was unclear whether present trends in motor contracting might be regarded as lasting. There remain, under the 1980 statute and implementing regulations of the ICC, certain regulatory advantages to contract service—in particular the greater freedom of carriers and shippers to negotiate with one another, and the opportunity to keep their dealings secret from competitors. The extent to which this may have skewed shippers' and carriers' decisions respecting the form of their relationships must be considered an imponderable until (and unless) these residual distinctions have been removed by still further deregulation.

Household Goods Movers

Although most household goods movers are motor carriers, and thus fall under that general heading (the balance are freight forwarders), the highly specialized nature of their service warrants separate consideration—a distinction that is also reflected in the law.

Unlike all other freight carriers, household goods movers

alone are regularly employed by consumers (*i.e.,* the general public). That lends this sector of the industry an extraordinary political sensitivity, which Congress felt impelled to address when it dealt with motor carrier (and forwarder) deregulation. Thus, although movers are generally accorded the same freedoms as other truckers and forwarders by the Motor Carrier Act of 1980, they are also the subject of their own separate statute, the consumer-protection-oriented Household Goods Transportation Act, enacted in the same year.

Respecting entry, however, movers were treated little differently by the 1980 deregulatory legislation than were motor carriers; and the impact on the industry has been roughly the same in terms of marketplace population. There are, that is, a great many more household goods movers competing for traffic today than there were in the pre-deregulation period.

At the same time, it is noteworthy that most of these "new" entrants are *not* new to the business of hauling household goods. Given its extraordinary marketing problems caused by the fact that it deals with consumers rather than industrial shippers, the household goods moving business is not an attractive field for the inexperienced entrant. But because of its peculiar structure, the industry already had a rich lode of potential new competitors within its own confines—those companies that had previously served as "agents" for the larger, regulation-authorized van lines.

The agency structure is unique to the household goods transportation sector. Most movers operate on a two-tiered basis; the "mother" carrier assumes responsibility for all aspects of marketing, plus certain other overhead administrative activities (including holding the requisite regulatory operating rights), but actual operations are performed through the offices of a network of local companies that represent these van lines as agents. Prior to deregulation, most of the agents handled only a small proportion of independent business (mostly within local areas where there was no economic regulation), with the bulk of their revenues deriving from their relationships with the major van lines they represented. Since 1980, however, a good many agents have opted to enter the market on their own, either in addition to or instead of continuing their agency status.

As of the date this was written, the influx of such ex-agents into the marketplace had relatively little effect on the upper echelons of the household goods transportation industry. The industry was always much more concentrated than other sectors of motor carriage, with a relatively few major carriers dominating the market; furthermore, the non-union status of these carriers (with their reliance on local agents and truck owner-operators to provide the actual service) made them less vulnerable to wage-cutting competition than carriers of other types of freight. In addition, the extraordinary marketing difficulties and requirements of a consumer-oriented industry such as household goods moving are a major obstacle to new entrants attaining significant market shares in this sector. A list of the top dozen or so movers of 1985, for example, contains mostly the same names as a similar list of five or ten years earlier.

Nevertheless, the much larger number of small competitors promises to have long-term effects on the industry's configuration. First, the new potential for local van lines to abandon their agency status in favor of operating their own moving businesses is creating considerable disruption in the large companies' agency structures. Agency turnover among the movers has considerably increased since 1980, and many of the larger agency-structured carriers have had to accept (albeit grudgingly) a new dualized status on the part of some of their agents, under which the mover is actually competing for certain traffic with its own agent.* As more agents shift the focus of their operations increasingly to running their own businesses, these effects can only continue to produce problems.

Not all household goods traffic is hauled directly on behalf of

---

*Although these dualized arrangements had not been subjected to serious legal challenge as of the time this was written, their status under antitrust laws must be regarded as somewhat uncertain. Should the courts deem such situations to be anticompetitive (and thus illegal) at some future time, the disruptive effect on the industry's agency structure is potentially profound.

consumers. A large (and growing) number of major corporations arrange and pay for the household goods moves of their transferred employees and new hires. Unlike the individual consumer, such corporations are potentially repeat customers for the movers; and accordingly a substantial part of the movers' marketing strategies is geared toward wooing and retaining these "national accounts." This has resulted, in particular, in a new form of household goods transportation since the deregulatory legislation was enacted—contract carriage.

As was the case with railroads, contractual household goods transportation was legally prohibited prior to 1980; indeed, it was not until 1983 that the first such operation was authorized by the ICC (and subsequent—unsuccessful—court challenges of this decision required another year or more to complete). Since then, contract service has expanded markedly within the industry. It appears likely that, in the course of time, the great majority of national-account service will be performed under contract, with most common-carrier customers being individual consumers. The proportion of common to contract carriage in this sector will, of course, be governed more by corporate personnel policies than by any factor directly related to the moving industry; but the development of volume rate structures under these contracts is likely to encourage more companies to offer this benefit to their employees.

Notwithstanding their descriptive name, household goods movers handle traffic other than the furniture, personal effects, etc., of individuals and families who are changing their residences. This is the so-called "third proviso" traffic (identified after the statutory provision that allows them to haul it under their household goods operating authorities)—unusually delicate or fragile industrial products that require the same careful handling and so-called "air-ride" equipment in which the movers specialize. Most commonly this consists of electronic equipment and instruments; also in this category are shipments of artwork, exhibit materials and other such goods.

Pound for pound, this traffic is exceptionally profitable for the carriers; indeed, many derive a major portion of their net revenues from handling it. And unlike household goods traffic (where the bulk of the movements are crowded into the summer months so that families will not disrupt their children's schooling), third-proviso traffic is largely non-seasonal in nature, another very attractive feature from the carriers' viewpoint. Yet a third attraction is that, since this traffic is *not* consumer-oriented, the carriers are relieved of both their added marketing stresses and the special regulatory burdens of the Household Goods Transportation Act when they handle it.

Accordingly, this traffic has become a focal point for competition within the industry; it affords at least the potential to become the key opening by which the new, smaller entrants can solidify their market positions in order to aspire to eminence within their industry. Again, both common and contract service is involved, with contract carriage likely to eventually occupy a dominant position.

Brokers

The vastly intensified competitive situation in the motor carrier industry has created an opening into which has stepped a previously more-or-less inconsequential participant of the transportation industry—the broker.

Brokers are a familiar presence in many economic areas, as middlemen in bringing buyers and sellers together. The real-estate agent who helps an individual sell or buy his a home is a broker; so is the entrepreneur who puts together a commercial construction project, or packages a bond offering, or arranges refinancing for a troubled company. The broker contributes nothing tangible to the actual transaction; but by putting the right parties in touch with one another, he makes it possible for that transaction to take place.

In transportation the best-known form of brokerage is the travel agency, through which one can arrange for not only personal transportation but also hotel and rental-car reservations, entertainment programs, etc. But there is also the less-publicized freight transportation broker, whose profession is helping the shippers and carriers of freight get together. Since motor carriage is the sector of the transportation industry offering the greatest range of choice—and hence the sector most in need of brokerage to assist establishment of the shipper/carrier link-up—most often the freight transportation broker deals primarily or exclusively with that sector.*

In generalized terms, the broker arranges for, but does not himself either provide or take in-transit responsibility for, the transportation service. Within the confines of this broad definition, however, there are three traditional variations on the theme, and a fourth, post-deregulatory version that may ultimately outstrip all the others in importance.

(1) *The true motor carrier broker.* He approaches his function from the perspective of, and is compensated for his services by, the carrier; essentially he is a carrier sales agent, although he may represent many carriers in that capacity, even carriers in direct competition with one another.

Under regulatory standards incorporated in the Motor Carrier Act of 1935 and left in nominal force by the 1980 revision, the motor carrier broker who acts in this capacity must be governmentally licensed; a few states also maintain similar requirements for brokers dealing with intrastate trucking. The 1980 statute reduced licensing at the federal level to the merest of formalities; the broker need only demonstrate that he is "fit" to receive a license—a requirement that is usually met by the applicant's mere assertion that it is—and it will be granted one.

In light of this *de minimis* qualification for entry, the ranks of brokers increased dramatically during the post-1980 era. And as carriers scrambled to maintain and increase the raw volume of their traffic—a growing necessity in the face of diminished profit

*This is not to disparage brokerage involving other forms of transportation. Outside of motor carriage, the most common form of brokerage is concerned with international trade, where a multiplicity of parties—dealing not only with transportation but also other features of export/import commerce—may be involved and brokerage frequently facilitates the process. Such brokerage remains, but has changed little, in either importance or methodology, in recent times.

margins on each increment of business—they found a burgeoning demand for their services. Whereas the limited competitive picture prior to 1980 made them fairly unneeded in most markets (since there were relatively few carriers competing for the freight), the post-deregulatory growth in competition has given rise to an environment in which they play an increasingly important role from the carriers' perspective.

(2) *The shipper's agent.* The other side of brokerage, of course, is the party who does his work for the shipper, not the carrier (and is paid accordingly). In the jargon of transportation such a party is known not as a "broker" (although this is in fact the function he performs), but as a "shipper's agent."

Generally speaking, the shipper's agent serves as an extension of, or even a substitute for, the shipper's own traffic department. He is responsible for selecting the carrier to perform the actual transportation, often handles negotiations with carriers on the shipper's behalf, in many cases even arranges for pickup-and-delivery scheduling and other details of the transportation process. He will also frequently handle such ancillary functions as tracing undelivered shipments, filing loss-and-damage claims, even auditing and paying freight bills, for all of which he is compensated, on bases that vary widely from case to case, by the shipper who employs him.

The nomenclature "shipper's agent" is misleading in one key, legal respect. The term "agent" is used in the law to designate one who is empowered to act for someone else, and for whose actions that "someone else"—the principal—bears full legal responsibility. The "shipper's agent," on the other hand, is actually an independent contractor insofar as the law is concerned. The distinction here is that, to the extent it agrees to certain arrangements and accommodations with the shipper's agent, the carrier may *not* hold the shipper responsible should the agent default; for instance, if the carrier agrees to accept payment of its freight charges from the shipper's agent and the agent for any reason fails or neglects to pass on to the carrier the money he has received for this purpose from his customer (the shipper), the law says the carrier may not demand payment from the shipper.

A good many true brokers also have *alter egos* as shipper's agents, and vice versa. This can create obvious conflicts of interest, since the broker/agent obviously cannot represent the best interests of both sides of the same transaction. Anyone using the services of any broker/shipper's agent should therefore make certain the party involved is not thus straddling the fence with regard to the transaction(s) in question.

(3) *The freight consolidator.* His role is essentially the same as that of the shipper's agent, with one important addition: He deals principally or exclusively in low-volume shipments which he consolidates into larger loads before tendering them to carriers for transportation.

The term "low-volume" should not necessarily be taken to mean shipments of just a few hundred or a few thousand pounds, although this is indeed the most common type of shipment handled by such operators. Consolidation-related savings are available in any context in which carriers can achieve operational economies of scale. In the past one form of consolidator was the "marriage-car broker," whose business entailed matching single trailerloads of freight into pairs before tendering them to railroads for trailer-on-flat-car service. Since each rail flatcar accommodates two trailers, railroads offered substantially reduced rates for such trailer pairs as against what they charged for single-trailer shipments. Most railroads now do their own trailer-matching and have eliminated the rate differentials that gave rise to such marriage-car brokerage; but this illustration does serve to point up that consolidation opportunities are not always limited to extremely small shipments.

The consolidator must obviously play a much more active role in the physical transportation process than other types of brokers. In general he will maintain some type of terminal or warehouse where he receives the freight and physically aggregates it for onward movement; many also have their own truck fleets to handle pickup operations within the local area (and will also probably provide local, and therefore unregulated, point-to-point truck service within the same area).

A closely related type of operation, often performed by the same party, is that of the "freight distribution agent." Just as the consolidator amalgamates shipments for outbound movement at the origin end of the line-haul movement, so the distribution agent provides "break-bulk" service for incoming consolidated shipments at the destination end. Again, the distribution agent will have a facility through which the goods are handled, and a fleet of local vehicles (either of his own or through a relationship with a carrier specializing in such operations) with which to make deliveries within his area of service.

Most consolidators/distribution agents perform these services in conjunction with other economic activities for which they use the same physical facilities. Most commonly they are also commercial warehousemen, although occasionally they will be involved in for-hire transportation services of their own in addition to their brokerage-type activities.

(4) *The broker/carrier amalgam.* The newest (entirely post-1980) wrinkle in motor carrier brokerage is the broker who is affiliated with—may even *be,* under a different corporate identity—a motor carrier.

There are a variety of reasons underlying creation of these dualized entities, the strongest of which is the opportunity for regulatory avoidance* such an arrangement affords.

Such entities take advantage of the reality that, under the law and regulations, a broker's relations with its customers are not subject to any form of government regulation, and (2) the relationship of a contract carrier with its shipper are likewise *de facto* unregulated. The broker thus serves as the "front man" in such a combination, handling all dealings with the shipper; and the carrier half provides the transportation under contractual arrangements with its affiliated/*alter ego* broker.

It should be evident that such arrangements demand careful monitoring by the shipper. Being affiliated with a carrier, the broker half of the partnership will have a strong predilection for

*In the sense it is used here, "regulatory avoidance" is comparable with tax avoidance—a wholly legal means of avoiding regulatory burdens that would otherwise be imposed through operation of the law and/or implementing government regulations (as distinguished from regulatory/tax *evasion,* which is used to designate an *il*legal means of accomplishing the same thing).

employing the services of its affiliated carrier, even if such services are more costly (to the shipper) than alternative means of transportation. On the other hand, a shipper with good negotiating capabilities will usually be able to secure highly favorable arrangements from such combines, since regulatory impedimenta do not interfere with the one-to-one negotiations the parties may conduct.

Regardless of the type of brokerage involved, anyone making use of a broker, in any guise, should be aware that ethical standards are not always impeccable in this sector of the industry. Most brokers, like most people, are honest; but the open-ended entry policy of the post-1980 period has attracted a few individuals who lack this important quality. There have been a number of disturbing tales of individuals who set up shop as brokers, operate for a few months on the basis of paying carriers out of monies they have previously collected from shippers, and then abruptly vanish with considerable sums of money they have been paid by their shipper customers but have *not* passed on to the carriers.

Well aware of the black eye such incidents are giving the industry (some shippers and carriers are no longer willing to work with brokers because of past experiences of this nature), the Transportation Brokers Conference of America has established a strong ethical program to weed out such problems. The most concrete part of this program is a "bad apple" file, in which the TBCA identifies brokers that have been the subject of verified complaints of unethical behavior. Because of the need to protect oneself against possible problems such as those described here, the TBCA program is worth considering by anyone who proposes to do any substantial volume of business with a freight transportation broker.

Shipper Associations

Non-profit shipper associations enjoyed a considerable vogue in the 1970's, which has to some degree extended into the

1980's. The likelihood, however, is that this sector of the transportation industry has peaked in terms of growth, and is likely to attain no greater role in the future save perhaps in one fairly specialized area.

Essentially the shipper association is a membership-structured shipper's agent (*see the discussion of brokers above).* Like the independent shipper's agent, it arranges for, but does not actually provide, the transportation of goods; but it does so only on behalf of its members. As the "non-profit" designation suggests, any net revenues from the operation are divided among the members on a periodic basis in proportion to their participation, rather than being retained by the independent contractor.

Prior to 1980 shipper associations were especially attractive because of their entirely regulation-exempt status. But deregulation, and the ensuing rapid expansion of (and consequently increased competition in) the brokerage sector, have diminished this attraction; there is simply less benefit to be obtained from, and therefore less reason to assume, the risks of membership in a shipper association.

The risks are twofold. First, of course, there are those attendant to membership in any industrial organization; if the organization flounders financially, the members may be assessed additional and unexpected charges to make up the deficit. Where the organization is non-economic in nature (a trade association, etc.), the members have, of course, the option of dissolving it; but in the case of economically oriented groups such as shipper associations, this may prove an equally costly alternative.

Like most independent shipper's agents, the shipper association normally works by collecting money from the individual shippers (its members) to cover freight charges incurred on their particular traffic, and then assuming direct responsibility for payment of the carriers' charges. If the shipper's agent fails to pay the carriers due to economic problems, those carriers, as already discussed, have no recourse to the shippers themselves. On the other hand, if the shipper association is in default to the carriers, the law *does* allow the carriers to bill the individual shippers—bills for which they are legally liable *even if they have already paid the association.*

In other words, a shipper unfortunate enough to have membership in a shipper association which becomes insolvent may wind up having to pay some of his freight bills twice—once to the association, and then a second time to the carrier. This is more than mere hypothesizing; such problems have actually occurred in a number of cases. Many shippers have accordingly become more cautious about shipper association membership, and, especially considering the competition-induced reductions in brokerage and shippers' agent charges, it does not appear especially likely that there will be significant further growth in this market segment.

In one area, however, the deregulatory legislation has led to expansion of the shipper association sector—associations concerned with the compensation they receive from railroads for providing their own rail cars (either owned or leased). The Staggers Rail Act for the first time permitted such organizations to negotiate collectively with the railroads on this question; although they must first secure antitrust immunity from the ICC, this has proved, at least in the early years following enactment of the law, not to be a major problem.

Previously the shippers had been reliant on regulatory protection in this area. Since antitrust restraints largely barred such collective negotiations, the railroads were in a position to decide unilaterally what allowances they would pay for shipper-furnished equipment, and dissatisfied shippers could only appeal the carriers' decisions to regulatory authorities. The Staggers Act removed this regulatory protection; but recognizing that in most instances no individual shipper would have adequate bargaining leverage to negotiate for these allowances individually, Congress at the same time relaxed the constraints on collective action.

As a result, a number of new associations have been formed to deal with the carriers in this area. It is worth noting that in many instances such associations have succeeded in obtaining better arrangements than they had previously been able to secure by regulatory intervention. While such negotiations will obviously not be an everyday occurrence, the continuance of most such associations as "watchdogs" seems fairly well assured.*

Airlines

Air transportation underwent its deregulatory rite of passage both earlier and more comprehensively than did other modes.

The history of commercial aviation in this country exhibits striking similarities to that of railroading, especially as respects government involvement during its early developmental stages. Both modes required a heavy government assist to "get off the ground" (as it were)—railroads in the form of the massive government land grants of the 19th century, airlines in terms of the air traffic control system and airport facilities. And both paid a high price, in economic regulatory terms, for that assist.

Government aid in any form rarely comes without strings attached. For the airlines, as for the railroads, the strings took the form of absolute government control over virtually every aspect of their operations, including particularly market entry and exit. No air carrier was permitted to commence or cease any service without the say-so of the governing regulatory body, the Civil Aeronautics Board.

For four decades conventional wisdom held that such tight regulatory control was required by the "public interest." But by the mid-1970's this notion had come under serious challenge; and in 1978 the airline sector became the first U.S. transportation mode to have its economic regulatory strings cut. (Nominally air deregulation, unlike that of other modes, was to be accomplished on a phased schedule; and in fact the CAB remained in existence as an agency until the mid-1980's. But this phasing was heavily "front-end-loaded," especially as to entry and exit, so that effective regulatory controls had disappeared within a matter of a year or two after enactment of the deregulatory Aviation Act.)

Unlike for other modes, economic deregulation for airlines was total; save in the safety area (which in any event was always outside the CAB's purview, being governed by the Federal Aviation Administration), government abjured all controls over the carriers' market decision-making. In terms of the configuration of the marketplace this produced two major results.

First, allowing for the much greater capital-intensivity of the airline business, deregulation had much the same effect on air transportation as it did on motor carriage. That is, there was an explosion (relatively speaking) of competition, with existing carriers quickly moving to expand their market coverage and a significant number of new entrants intruding into previously protected markets.

Airline competition has, however, been much more narrowly focused than competition in any other transportation sector. Because airlines derive the bulk of their revenues from hauling passengers, not freight, the bulk of the competitive pressure has focused on a relatively small number of high-volume markets where substantial personal travel takes place. These markets quickly became—and remained—seriously overcrowded: coast-to-coast service between key Atlantic and Pacific origins/destinations, service to and from the vacation resorts of Hawaii and Florida, "corridor" service in the northeast, the midwest, the southwest and the west coast, and the like. Meantime, smaller markets to and from which volume was less heavy and/or predictable suffered overall losses in service (*see below).*

The airlines were more fortunate than the motor carriers in one respect; because their deregulatory legislation was enacted two years earlier, they did not have to confront the 1981-83 recession just as they were having their first experience with regulatory freedom. Even so, the abruptly heightened competitive pressures led, much as it has (and is) with motor carriage, to an industry "shake-out" that, eight years after the Aviation Act became law, was still far from complete.

And unionization has also played a somewhat smaller role in airline competition than it has among motor carriers. This is partly because labor accounts for a considerably smaller percentage of total expenses for airlines than for truckers, and also because airline unions have been more willing to reach compromises, including the much-publicized "give-backs" several financially stressed carriers have been able to secure.

This is not to say union/non-union economic differentials have been inconsequential; certainly the advantages of lower wage scales and less restrictive work rules have been instrumental in the successes of such deregulation-spawned entrants as People Express, New York Air, etc. And the managerial decision by Continental Airlines to declare bankruptcy in order, principally, to escape its union contracts emphasizes that the costs of unionization are not competitively insignificant (Continental resumed operations, under a trusteeship and without union representation for its employees, within a matter of days). Overall, however, this aspect of the new market competition among air carriers has not been nearly so important as it was (and is) in post-deregulatory trucking.

Essentially what all of this means is that competitive pressures in the airline industry have been appreciably different, under deregulation, than among carriers of other modes, and that accordingly the ebb and flow of carriers out of and into markets (or solvency) has been based on different criteria. The net effect, however, has been much the same—established carriers under pressure both from one another and from a good many new entrants into the industry, and the new entrants themselves subject to the same competitive problems, with an overall instability to the marketplace.

The second major deregulatory result was unique to the airlines: the rapid development of an industrywide "hub-and-spoke" market configuration. Whatever the future comings and goings of carriers in the marketplace, this seems likely to be a permanent fixture.

The dependence of the passenger-oriented carriers on larger population centers has already been discussed. The reasons for it are obvious. As with any mode of transportation, the bulk of airline costs are concerned with the operation of the unit of equipment from Point A to Point B. Once the decision has been taken to operate that unit, each additional increment of revenue traffic (passenger or freight) handled is almost a pure financial plus; there are only the most minimal incremental costs associated with the additional traffic.

Railroads and motor carriers are able to sell their capacity in relatively large increments; often the decision to operate a unit of equipment and the full utilization of the capacity thus made available are coincidental. Airlines, on the other hand, sell their capacity in discrete units, and it generally takes quite a few of those units (there are, for example, as many as 400 seats on an individual airplane, each of which will usually be sold independently) to bring utilization to full, or even economically acceptable, levels. Airlines thus must be far more concerned with the notion of "load factors" (proportion of revenue-producing capacity to total available capacity) than are carriers of other modes, since it is much more likely for them to be unable to fill capacity than for the other carriers. This has led them, in logical consequence, to focus principally on high-population areas (relative, of course, to the capacity of the equipment they operate), where their chances of filling their available passenger seats are accordingly greater.

Thus, just as it could be said in ancient times that "all roads lead to Rome," so do all routes of any given airline tend to originate and terminate at just one or a few major population centers. In order to travel between any two points other than these high-population "hubs," the passenger (or the freight) must normally take one flight into the hub and another back out; or, if the origin and destination points are located in different regions of the country, a two-connection movement will be required (one flight from origin to local hub, a second "inter-hub" flight, and a third from the other hub on to actual destination).

This is vastly different from the way airline routes were structured prior to the 1978 deregulatory statute. Back then, with the CAB exercising rigorous control over market entry and exit, point-to-point flights between even secondary origins/destinations were far more frequent. To be sure, average load factors (both passenger and cargo) for such flights tended to be low; but the competitive restraint exercised by the regulatory body allowed airlines to increase rates and fares to compensate, and, furthermore, mandated considerable cross-subsidization of carrier operations. It was only the advent of deregulation that placed an economic premium on improved load factors and so dictated the hub-and-spoke system that has over the intervening years become so commonplace.

Naturally, competition has largely followed the same path. The spokes of the various carriers' operations, involving service to or from secondary markets at one terminus, have lured few entrants; indeed, a substantial proportion of secondary markets have lost at least some service since 1978, and even fewer are served by more than one or two carriers. On the other hand, hub-to-hub operations—where there are sizeable population centers at both ends of the route—have been competitive magnets, attracting so many carriers that there are often inadequate airport facilities to accommodate them all.

This situation has led to a curiously inverted form of cross-subsidization among most carriers, in which the bulk of their traffic is handled on a hub-to-hub basis but the bulk of their revenues come from their service along the spokes. To illustrate how this works, consider that, as of the time this was written, it cost the same for an individual passenger to fly between Atlanta and Montgomery, AL—a distance of less than 200 miles—as it did to fly cross-country between New York and San Francisco. Charges for cargo service, although not always so widely varied, tended to follow the same pattern.

In a third area—the delineation between common and contract service—airline deregulation has had relatively little impact. The reason is (again) obvious: Few air cargo shippers have sufficient volume to warrant, from either their own perspective or the airline's, establishment of a contractual relationship. For the most part (there are always exceptions), air freight shipments, being far more costly than by any other transportation mode, are made by shippers on an occasional rather than a regular, predictable basis. Since deregulation (especially when juxtaposed with deregulation of other modes at more or less the same time) has made little difference in the relative freight charges from one mode to the next, it has not served to significantly alter shipping patterns in this respect; so the shippers with sufficient volume to support contracts tend to be the same as those of the pre-1978 era, and if they have enough traffic for contracts now they likewise did then.

Domestic Water Carriers

These are the barge lines that ply the Great Lakes, the Mississippi River System and the coast-hugging Inland Waterway System. According to the statistical compilation *Transportation in America**, these carriers carry a very respectable 18 percent of all domestic freight tonnage—and, of even greater significance, account for more ton-miles than any other single mode of U.S. transportation (a "ton-mile" is one ton of freight moved one mile).

Deregulation has had much less impact on these carriers than on those of other modes, in part because a substantial proportion of their service was not regulated prior to 1980. Under the so-called "three-bulk-commodity exemption," any barge movement whose cargo consisted of no more than three different commodities, each moving in bulk (rather than packaged), was always exempt from government economic regulation. A substantial volume of barge traffic, especially on the Great Lakes and the river system, has always been comprised of such commodities as coal, iron ore, grain and the like, and such movements generally qualified for the exemption. Deregulation made, of course, no change in the status of these movements.

In addition, because it is by nature high-volume, a great deal of pre-1980 barge traffic was contractual in nature. The relative ease of securing contractual operating authority even in the era of stringent regulation has already been discussed respecting motor carriage, and the situation was no different for the barge lines. Entry restrictions, accordingly, exerted considerably less of a

*Third ed., March, 1985. Transportation Policy Associates, P. O. Box 33633, Washington, DC 20033.

restraining effect on competition here than in other transportation sectors, and their relaxation (by clauses in the 1980 Motor Carrier Act that were fundamentally the same as for trucking) was therefore of less importance.

Thus, deregulation has not produced in domestic water carriage the competitive explosion that has characterized motor and air transportation; nor has it led to massive reductions in the scope of service such as has been the case in railroading. Basically, like the waters on which they operate, barge carriers have "just kept rollin' along" in more or less the same vein they always did.

The one thing that may ultimately affect the role of barge transportation in the future marketplace is the potential for these carriers to combine with other modes (*see below*). It is possible that such combinations—especially where they involve railroads, which have always been the barge lines' principal competitors as well as their main connections for traffic the originates at or is destined to points not directly on navigable waterways—may ultimately result in substantial shifts in water traffic patterns. As of the time this book was prepared, however, only one such combination had been formalized, and the impact of it and any possible future transactions of the same nature was uncertain.

Freight Forwarders

As it did for truckers, the Motor Carrier Act of 1980 likewise largely deregulated domestic surface freight forwarding (to roughly the same extent it did motor and domestic water carriers); in addition, it removed a restriction that had theretofore barred forwarders from establishing contractual relationships with the "underlying" rail and motor carriers that actually perform the transportation services the forwarders offer to the shipping public. And the Aviation Act of 1978 treated air freight forwarders to the same total deregulation it accorded airlines.

Forwarders are (although they probably will consider this a derogatory definition) basically a hybrid between brokers and actual transportation carriers. Like brokers, they arrange for, but do not actually themselves operate, transportation services; but like carriers they assume full responsibility and (of particular significance) loss-and-damage liability for the freight while it is in transit, which brokers do not. Actually, a nomenclature more clearly descriptive of their services is that employed by the maritime industry respecting international trade—"Non-Vessel-Operating Common Carrier" (NVOCC).

The primary function of the freight forwarder is to help shippers and carriers alike attain the economies of scale available from increased volume. As already discussed, all but a fraction of any carrier's operating costs are associated with running a unit of equipment; little, if any, significant cost differences relate to the volume of freight loaded aboard that unit. This has already led railroads (in the 1960's) to discontinue all less-than-carload (LCL) service; and it is reflected in the trucking industry by rate scales that assess on smaller less-than-truckload (LTL) shipments rates two or three times (or more) higher, on a per-hundredweight basis, than apply as on truckload (TL) movements. (Air service, by inherent nature more geared to smaller shipments, exacts less economic penalties on small shipments, but rate differentials are still quite substantial.)

The forwarder accumulates smaller shipments from many shippers, consolidates them for transportation as volume lots, and receives his compensation by the difference between the small-shipment rates he assesses those shippers and the higher-volume rates for which he is able to engage the services of equipment-operating carriers. In this way forwarders are able to make available rail service to shippers who lack adequate shipment volumes to actually fill a rail car, to afford to shippers the generally superior TL service of motor carriers instead of less reliable LTL service, and in the process to earn a profit.

In its initial phases, deregulation has done little to change the surface forwarding industry. Entry restraints in the forwarder industry were never a critical factor in limiting competition in this mode, so that their removal has had little competitive impact. And the new statutory permission for forwarders to contract with rail and motor carriers has been considerably vitiated by the general unwillingness of these carriers—who commonly regard forwarders as their competitors in the marketplace—to agree to favorable contractual arrangements.

The air forwarder industry, on the other hand, has experienced some degree of growth in the post-deregulation era. This is not altogether surprising, since air freight tends to be extremely low-volume; furthermore, airlines have dedicated the bulk of their competitive efforts toward passenger service, and have thus had more reason to be grateful for the offices of forwarders in helping them fill their cargo capacity. Fairly obviously, if the airline industry ever decided to make a concentrated effort to win business away from the forwarders it could make substantial inroads in their business; but the likelihood of this happening among the primarily passenger-oriented carriers seemed, as of this writing, fairly small for the foreseeable future.

Even so, forwarding continues to account for only a tiny proportion of freight movements. Tonnage and ton-mile figures are not available, but for 1984 (the most recent year for which figures were available when this was written) forwarders accounted for only about a quarter of 1% of the nation's total freight bill.*

Pipelines

In terms of tonnage and ton-miles, pipeline carriers compare favorably with the other major modes of transportation.† They have, however, been little affected by transport deregulation, which in any case did not, for the most part, affect them.

**Transportation in America, op cit.*

†*Transportation in America, op cit.,* reports information only for oil pipelines. In 1984 (the latest year for which figures were available when this book was written), oil pipelines were estimated to account for 17.8% of the total intercity freight tonnage (compared with 37.1% for truck, 27.1% for rail, 18.0% for barge and 0.1% for air), and 18.8% of all ton-miles (34.4% for barge, 28.2% for rail, 18.4% for truck and 0.2% for air).

The main traffic moved by pipelines consists of petroleum and/or its products. As such, oil pipelines tend to be owned and operated by the petroleum industry, which is also their only customer. Recognizing this, Congress removed oil pipelines from the economic regulatory structure in the mid-1960's; and their status has not been altered by subsequent legislation.

Additionally, there is considerable pipeline movement of natural gas. Again, this is petroleum-related; moreover, natural gas pipelines have never been considered part of the transportation industry, but have rather been treated under laws concerned with the production and distribution of energy.

A very small volume of slurrified coal—coal that has been pulverized into a powder and then mixed with water to make it "flowable"—is moved by pipeline. During the course of the deregulatory movement of the late 1970's and early 1980's there was a good deal of consideration given to expansion of coal slurry pipelines, a notion that seemed especially attractive when the so-called "oil crisis" of the 1970's radically escalated prices of petroleum-based energy and threatened continued availability of oil imports (thus increasing the relative importance of domestic coal as an energy resource).

The building of such pipelines could not be accomplished, however, without use of the so-called "right of eminent domain" to require landowners to let the pipelines be buried under their property; and that right could only be conferred by governmental action. The railroad industry, which has a *de facto* virtual monopoly on U.S. coal traffic, lobbied intensely against such action; and as of this writing it appeared that, for the moment at least, these efforts had succeeded in preventing the construction of a new coal slurry pipeline network.

Intermodal Transportation

Deregulation has eliminated many legal impedimenta to the development of intermodal transportation service; but, although this is leading to some degree of growth, the ultimate role of intermodalism in the transportation marketplace remains uncertain.

Regulatory law—developed as it was decades ago, long before the technology to support serious intermodal operations was in place—took no account of this type of service. It was based on the premise that the various transportation modes were operationally unrelated by their inherent nature; and in order to prevent carriers of any one mode from asserting monopolistic influences on the transportation market by controlling intermodal competition, it essentially required that ownership and operation of each of the modes remain separate. (At a time when other modes were in their relative infancy, the well-established railroads were the main targets of this legislative restriction.)

Over the course of time two things took place. First, such developments as rail trailer-on-flat-car (TOFC, or "piggyback") service, maritime containerization and the like came "on-line" in transportation due to technological advances. Second, the relative economic strengths of the various transportation modes became more equalized. In short, it became operationally desirable for the carriers of various modes to be able to work in closer coordination with one another, even to become economically affiliated; and the aboriginal reasons for restricting such combinations diminished in importance.

Increasingly, too, it became evident that this outmoded mandatory segregation of the modes was holding back potentially beneficial marketplace developments. The growth of railroad TOFC traffic was seriously impeded by these strictly applied regulatory limitations; so were the various "land-bridge" (and the variant "mini-bridge" and "micro-bridge") operations involving ocean and domestic service combinations. (This latter situation was especially exacerbated by the fact that the two modes were regulated by different agencies—the maritime carriers by the Federal Maritime Commission, their domestic connections by the Interstate Commerce Commission—each of which exhibited typical bureaucratic jealousy of any encroachment by the other on what it perceived as its own jurisdictional preserve.)

As in essence a by-blow of its deregulatory measures, Congress has largely stripped away the modal-separation policies and requirements of the law. Intermodal coordination, affiliation and ownership are now generally deemed a matter of marketplace decision-making (subject only to the restraints of antitrust laws), so that intermodalism may proceed legally unencumbered.

During the early post-deregulatory years, however, the decades of regulation-enforced modal apartheid continued to exhibit a strong residual effect. Steeped in the the disciplines, the traditions and (most of all) the *esprit* of their own modes, carrier managements shied strongly away from combining, or even cooperating, with those of other modes. Such relationships as did exist between carriers of different modes continued to have very much an arms'-length flavor; and more often there were no relationships at all, so that the shipper of intermodal traffic would often be obliged to make separate arrangements with each participating carrier.

By the mid-1980's, however, there were a few signs that these managerial attitudes might be crumbling in the face of economic reality. In particular, piggyback traffic grew dramatically; in several instances railroads acquired motor carriers (and in one case a barge line) to complement their rail services; and there was also a considerable spurt in the various forms of containerized "bridge" operations.

It is obvious that further developments along these lines are to be expected; but just how far this intermodal movement will go is unclear. Even the changes that have already taken place should not be deemed an unmitigated endorsement of intermodalism by the carrier community; for example, the big TOFC spurt has come mostly at the expense of rail boxcar service, allowing the railroads to retain the traffic and revenues while reducing a type of service (boxcar) most of them would prefer to abandon anyway. There are also serious questions as to whether intermodal combinations will produce significant economic benefits, since few operational economies of scale (or otherwise) are readily visible.

It would appear reasonable to anticipate that, in the not-too-distant future, a shipper whose traffic must move via more than one mode will be able to accomplish this on a single-factor basis (one set of shipping documents, one freight bill, etc.). It is also likely that there will be some further movement toward intermodal acquisitions or mergers. Beyond these fairly obvious forecasts, however, it would be a mistake to be too "bullish" on the future of such intermodal trends; the marketplace has yet to reach its ultimate consensus on the future of this movement.

Private Carriage

No discussion of the transportation marketplace would be complete without mention of private carriage.

The term "private carriage" must be considered to include the implicit modifier "motor," since virtually all private movements of freight are by this mode. To be sure, a handful of firms operate their own short-line railroads, their own aircraft, their own water vessels; but the aggregate volume of freight moving by all of these is minuscule. By comparison (although no statistics are available), it is generally conceded that a very substantial proportion of truck freight moves in vehicles owned and/or operated by the shippers or receivers of that freight.

As of when this book was written, it was impossible to assess the overall effects of deregulation on private carriage. The Motor Carrier Act of 1980, and implementing policies subsequently adopted by the ICC, did much to eliminate decades-old regulatory obstacles aimed at discouraging this form of transportation; but the more rigorously competitive environment in the post-deregulation for-hire marketplace likewise reduced the incentive for shippers and receivers to provide their own transport service. It was uncertain (and likely to remain so for years to come) which of these countervailing influences, if either, would predominate.

As already mentioned, in the halcyon days of regulation both the law and regulatory policies were aimed at fostering common carriage as the primary form of U.S. transportation, and discouraging other forms (such as contract and private carriage) which might weaken the common carrier network by depriving it of traffic and revenues. This attitude led to a number of restrictions on what the law deemed *bona fide* "private" carriage within the scope of the regulatory exemption extended to such service—

restrictions that enforced on private carriers extensive economic inefficiencies and thereby (it was hoped) would discourage shippers from electing this alternative.

Nevertheless, private carriage exhibited robust growth throughout this span. The then-prevalent rationale for this phenomenon was that, despite its regulation-dictated inefficiencies, private carriage could afford shippers better transportation service and/or lower costs than the market-protected, competition-poor common carrier sector of that era.

In fact, this was always a vast oversimplification. Transportation service and rate factors are only a part of the reason a shipper may be induced to establish its own private carriage operation, and not necessarily the most important part at that. Corporate pride, a desire to maintain a marketplace "presence" among customers, competitive pressures within the shipper's own industry and a variety of other factors also play important roles in this decision. If there was ever any doubt that this is true, private carriage's continued display of strength in the post-deregulatory period—when the service/rate problems that supposedly led to its growth have presumably been removed from the much more competitive for-hire transportation market—should dispel it.*

The Motor Carrier Act of 1980 made one substantive change in the restrictions previously applicable to private trucking; recognizing the trend in U.S. corporate development that has led to establishment of corporate "families" of affiliated concerns, it removed the limitation that private carriage could only be performed within the confines of an individual company to qualify for its regulatory exemption. This allowed any such "family" of corporations—provided that they are 100% co-owned—to share the services of a single private carrier, rather than obliging each

---

*For a more extensive discussion of the variety of reasons that underlie corporate decisions to operate private carriers—and, of course, a fuller discussion of this form of transportation in general—see the author's *Practical Handbook of Private Trucking* (Washington, DC: The Traffic Service Corp., 1983).

to operate its own, and thereby gave them the opportunity to realize the economies of scale available from a broader operational base.

Post-1980 policy changes by the ICC have gone still further in relaxing the historic limitations concerning what does and what does not constitute private carriage for purposes of the regulatory exemption. In particular, limitations on private carriers' freedom to lease equipment and drivers from outside sources have been reduced or eliminated; private carriers have been accorded permission to lease their equipment and drivers to others (including regulated for-hire truckers) to avoid underutilization; and the door has been opened at least part-way for non-affiliated private-carrier-operating shippers to enter into cooperative ventures in which they share the services of a single fleet and roster of drivers.

Although (as noted) no overall trend has emerged, the post-1980 changes have created a fair amount of movement within the private carriage sector of the industry. Some shippers, spurred by the opportunities they perceived in the more competitive post-deregulatory for-hire marketplace, have curtailed or discontinued their private carrier operations; others, with little or no previous involvement in private carriage because of the economic disincentives previously imposed by the regulatory structure, have begun or expanded such operations. Whatever the ultimate verdict of history, however, it is evident that private carriage will continue to play a role of at least some significance in the transportation market.

3

REGULATION AND RATES TODAY

"Anything Goes."

Those whose memory goes back to the 1940's—or those who are afficionados of the music and/or motion pictures of that era—will recognize this as the title of Cole Porter's award-winning Broadway musical play and motion picture. More pertinent to the contemporary transportation industry, however, it's also the rule of thumb applicable to shipper-carrier relations in the 1980's and, presumably, beyond.

Regulation, so strong a force in transportation for nearly a century—and so restrictive in what it did and did not allow in terms of transportation service and rates—has been relegated to a status of unimportance in today's industry. Even the limited controls left in place by Congress in the Staggers Rail Act, the Motor Carrier Act of 1980 and related statutes, have been largely reduced, *de facto,* to relative inconsequentiality by rulings of the Interstate Commerce Commission and/or the courts.* It is today almost exclusively the exigencies of the marketplace, and not the mandates of government, that dictate the terms of the relationship between any carrier and its shipper/customers.

This is eminently reasonable, in light of the basic removal of market entry constraints on the carriers. Since entry relaxation was intended to engender increased marketplace competition, it

*And it appeared, as of the time this was written, that further legislation would soon be enacted to extend and enlarge the scope of deregulation on a statutory basis.

would have been senseless for Congress to have taken this action while retaining regulatory controls over the manner in which carriers might compete. Or, to put it another way, if it makes sense to open up entry to enhance competitive opportunities, it must also be deemed sensible to trust the resulting competition to exert appropriate control over the carriers' ratemaking. Either way, entry deregulation without concomitant rate deregulation would be an obvious logical inconsistency; and Congress coupled them inextricably together in its legislative actions.

Carriers are today free to offer their shippers just about anything those shippers want in terms of both rates and service—which means, realistically, anything on which they and their shippers can mutually agree. Common rail, motor and water carriers, and domestic surface freight forwarders, must generally publish their rate and service offerings in tariffs filed with the ICC (*see below*—although there are limited regulatory exemptions that relieve them of even this obligation in some circumstances); air carriers and forwarders need not issue formal tariffs for any services; and all contractual forms of service are today regulation-free as respects both the form and the content of shipper-carrier agreements.

Common-Carrier Tariffs

Even the tariff-publishing constraint, where it applies, has been reduced to a minimum. First, the old 30-day rule—requiring that any tariff changes be published 30 days before they would be allowed to go into effect—has been eradicated. Railroads may today publish new or reduced rates (*i.e.,* any tariff provisions that either establish new services or cut the freight bills shippers pay for existing services) on just 10 days' notice; rate increases may be made effective after a 20-day notice period. Motor carriers may effectuate new or reduced rates on but a single day's notice, and need wait only seven working days after publication before rate increases go into effect.*

*The "publication" of a tariff, for purposes of computing the time, is

Second, regulatory constraints on what tariffs may say and how they may say it have been almost entirely scrubbed. As a matter of practicality, it is safe to say that virtually any agreement a shipper and carrier can negotiate may legally be published in the carrier's tariff. A few examples:

• *Discount tariffs.* For simplicity's sake, many carriers employ rate scales published in generic tariffs (often those published by rate bureaus—*see below),* but allow liberal percentage discounts off those rates. In some instances certain marketing-oriented conditions are attached to those discounts. For example, some carriers require shippers to apply in writing—usually in advance of making shipments—to be eligible to receive discounts; this may even be combined with provisions obliging shippers to pay freight bills computed on non-discounted rates, with discount "refunds" being paid back later. The objective of such provisions is, of course, to limit the discounts to the carrier's regular customers (who presumably have met the requisite conditions) and not extend them to casual shippers.

• *Volume and other incentives.* In constant search for shippers who can be counted on as regular revenue sources, carriers in many cases are conditioning eligibility for discounts on the shipper's tender of continuing volumes of traffic; if the shipper falls short of the volume requirement (computed on a periodic—monthly, quarterly, etc.—basis), the discount doesn't apply. Other discount conditions may be based on individual-shipment volume (such as multiple-pickup discounts on small-shipment traffic), the provision by the shipper of certain services (loading, unloading, etc.), and so forth. In still other instances carriers extend discounts for marketing-related purposes, as where discounts are offered to consignees who, when they purchase goods, specify transportation routing via those carriers. In this context it is worth noting that there is no requirement that discounts be

deemed to take place on the calendar day the new tariff matter is delivered to the ICC's Washington headquarters—which, since guards are on duty at the Commission 24 hours a day and are officially authorized to "receive" any such tariff publications, may extend to as late as 11:59 p.m. Once again, the foregoing applies as of the time this book was written.

paid only to the party responsible for the freight bill; thus, consignees may in some cases earn discount payments from carriers on prepaid shipments, shippers on collect shipments, etc. (*although see Chapter 5 for a further discussion of this subject).*

• *Named-shipper tariff provisions.* It is even legal for carriers to identify in their tariffs, by name, individual shippers to whom particular provisions apply. A carrier may thus maintain one rate-and-service package for Shipper A, another for Shipper B, and so forth—just as if it had separate contracts with each shipper (and even though no actual written contracts exist and contractual service is therefore not, in a formal sense, involved).

• *Incentives and penalties.* Tariffs may, at the carrier's option, spell out service standards together with economic incentives for meeting them and/or penalties for failing to do so. This is most commonly employed in connection with transit times: The carrier offers time-guaranteed service based on the date or time of pickup, receiving an extra percentage if it makes timely delivery and/or forfeiting a penalty (as much as 100% of its freight bill!) if it fails to. There are also numerous other service specifics that may be covered by incentive/penalty provisions.

• *"Simplified" tariffs.* Probably the best-known examples are the motor carrier ZIP-Code tariffs, which divide the nation into segments based on the U.S. Postal Service's three-digit ZIP Codes and establish standardized rates between each ZIP pair. Other simplified tariffs are based on other approaches; some simply assess a flat charge per mile between origin and destination (with, perhaps, a sliding scale based on distance), or otherwise eliminate the traditional complexities of transportation tariffs. Perhaps the ultimate in this vein are tariffs that establish flat charges, based on averages of a shipper's past traffic, for all shipments within broad weight and geographic ranges.

New rules promulgated by the ICC do much to encourage flexibility in ratemaking and tariff publication. One of the most potentially innovative regulatory changes of the post-deregulation era, for example, even allows carriers to incorporate advertising or promotional material in their published tariffs—allowing those tariffs to be transformed, at the carrier's option, into catalog-type publications of the same genre that makes mail-order sales operations so successful. The same changes also rescind virtually all other past restrictions concerning the format of carrier tariffs; as an indication of just how far thisreaches, carriers may today legally write their rates out on 3x5 file cards, irregularly shaped scraps of paper, etc., and file them just that way with the Commission as legal tariffs.

In addition to these and other such freedoms, current regulatory law also allows shippers and carriers to reach agreements over how tariffs are to be interpreted, without regard to previously binding legal standards in this area. The Commission has formally declared that "the issue of future tariff applicability may properly be the subject of negotiation between carriers and shippers," and added that "we encourage such negotiations." In theory, carriers are obliged to interpret their tariffs consistently for all shippers; that is, a carrier may not favor one shipper with a liberal reading of its tariff while interpreting the same or similar language differently for others. Since such "negotiated" tariff interpretations are essentially private bi-partite agreements between individual carriers and shippers, however, it is difficult to see how this limitation can be effectively enforced by either regulatory authorities (which have, in any event, shown little interest in doing so) or competing shippers and/or carriers.

For all the freedom and flexibility now permitted in tariff publication, however, one thing has not changed: Tariffs are still legally binding on both common carriers providing service subject to regulatory jurisdiction and their shippers. Such carriers must, by law, assess charges based strictly on what their tariffs say; and shippers are equally obliged to pay such charges.

During the first few years of transition to a deregulated transportation marketplace this requirement was often neglected, especially in the trucking sector. The 30-day tariff rule was still in effect at that time for motor carriers, and both the carriers and their shippers alike chafed at their inability to negotiate rates more promptly in the newly market-competitive environment. As a result, it was not uncommon for them to adopt their negotiated agreements as the basis for billing and payment even though those agreements differed from published tariffs. A device sometimes employed was for the carrier to agree to "protect" negotiated rates (*i.e.,* base its freight bills on such rates) during the interval between completion of the negotiations and effectiveness of the new tariff provisions.

It is important to recognize that, no matter what wording is used to describe it, *any* deviation from the tariffs of carriers still under legal obligation to publish them—in carrier billings, shipper payments or both—is strictly illegal. The so-called Elkins Act (the provisions of which are now incorporated in the recodified Interstate Commerce Act) establishes both civil and criminal penalties for carriers who offer or give, and shippers who solicit or accept, any "rebates or offsets" against tariff rates.*

In reality, the likelihood of prosecution for Elkins Act violations in today's deregulated environment is remote. The ICC, which has prime responsibility for enforcement of this statute, has dedicated very little of its resources to this area; only the most egregious and flagrant of violations have been prosecuted at all, and even these cases have been mostly settled by carriers and/or shippers merely agreeing to change their ways in the future. Others, including private parties (such as competing carriers or shippers), are nominally permitted to bring Elkins Act prosecutions too, but, again, this rarely happens.

---

*This statutory language, it should be noted, is quite broad enough to cover any non-monetary concessions as well. Thus, if a carrier provides extra increments of service above what its tariff specifies, fails to bill for demurrage and/or detention charges, allows extended credit periods, etc., both it and the shipper who receives these benefits are violating the law and are subject to the same penalties. However, it must be emphasized that the Elkins Act and other statutory provisions concerning the binding nature of tariffs apply only where the law requires publication of rates in tariff form. Some regulated carriers incorporate in their tariffs provisions that cover *non*-regulated services—motor transportation of agricultural commodities, rail piggyback service and the like. The ICC will accept such tariffs for filing; but because the service isn't subject to any tariff-publication requirement, these tariffs are not legally binding on either carrier or shipper under the Interstate Commerce Act.

As many shippers have learned to their dismay, however, the unlikelihood of prosecution does not mean these provisions of the law may be violated with impunity. Notwithstanding any negotiated agreement that it will charge lower rates, the carrier is always free (indeed, is legally obliged) to bill on the basis of what its published tariff specifies. In other words, the carrier can welch on the deal, and the shipper has no recourse. This, too, rarely happens, of course; but the problem surfaces with a will where carriers fall into bankruptcy—a not-uncommon phenomenon in the rigorously competitive transportation marketplace of today.

Courts overseeing the reorganization or liquidation of bankrupt carriers routinely appoint outside auditing/collection organizations to review past carrier billings and collect any undercharges. (Because of a peculiarity of bankruptcy law, by the way, these collection agencies are not limited by the usual three-year statute of limitations that applies to undercharges on regulated service, but may take action on shipments made as long as *five* years previous.) Not surprisingly, such audits are conducted in strict accordance with tariffs in effect as of the time of each shipment, and totally without regard to any illegal extra-tariff agreements. Many shippers have found themselves belatedly billed for undercharges in these circumstances, with no alternative but to pay.*

With regard to railroads, the problem can be less severe due to a peculiarity of the law. In effect, it is legal for railroads to negotiate *ex post facto* rates—rates applying to shipments moved before tariff publication, or even before the negotiation itself took place. ICC approval is required in such cases; but there

*Courts and the ICC have exhibited considerable sympathy for shippers in these circumstances, especially where it appeared that the shippers, as of the time the shipments were made, were ignorant that the agreed rate was not in fact formally published in the carrier's tariff. As of the time this book was written, however, this sympathy had not been translated into legally sound legal decisions relieving shippers from having to pay such tariff-based undercharges.

are two separate bases on which it may be sought, and in most instances it will be routinely granted. However, this applies *only* to rail transportation, and not to any other mode.

The simplest way of handling such situations is for the railroad to seek permission, under special ICC regulations, to waive collection of charges higher than the agreed level. A second approach is for the carrier to request that the past shipments in question be exempted from regulation, which relieves both them and their shippers from the requirement to stick to published tariffs. Both obviously require the carrier's willingness to charge less than its tariff states; there is no way for a shipper to enforce any extra-tariff agreement against an uncooperative carrier.

A third, still more recent option (adopted by the ICC in the waning months of 1985) allows railroads to *legally* "protect" contractual rates from the moment they are negotiated—even though the contract itself may not have been formally signed (and filed with the ICC; *see Chapter 8),* or even set down in writing. This is, again, limited to rail carriers and does not apply to motor or domestic water service or freight forwarding.

Reduced Protections for Shippers

In light of the freedom deregulation has conferred on shippers to choose the carriers whose services they want to use, and to negotiate effectively with those carriers, it should not be surprising that deregulatory laws and policies have also stripped away most of the shippers' erstwhile regulatory protections. If today's shipper cannot negotiate the rate/service package it wants in the competitive marketplace, its failure to achieve its objectives will not ordinarily be appealable to government regulators.

This is formally written into the Staggers Rail Act, which specifies that "a rail carrier. . . may establish any rate for transportation or other service provided by the carrier." The words "any rate" are meant to be taken literally; there are no regulatory controls whatever on most railroad ratemaking.

The exception to this rule comes into play only where a

railroad (or consortium of railroads participating in interline service) has "market dominance" over the traffic involved. A twofold test is employed in market-dominance determinations. First, there must be no "effective competition" for the traffic—either from other railroads or from carriers of other modes. In addition, where inbound traffic is concerned—that is, where the complaining party is the consignee—the Commission will also consider "source competition" (whether identical goods could be procured from other sources via other transportation routings) and "product competition" (whether different but comparable goods could be procured by means of alternative routings).

Even proof that there is indeed no effective transportation competition, under all of these tests, will not alone demonstrate the existence of market dominance. The law also provides that the rate levels must be abnormally high, generating revenues for the carrier (or, in a joint route, the participating carriers) equal to the rail industry's calculated "cost recovery percentage." This percentage, which is based on a comparison of revenues and carrier variable costs and involves some extremely complicated calculations, is "capped" by the Staggers Act at 180%—that is, revenues equalling 180% of variable costs. Since every computed cost recovery percentage since enactment of the Staggers Act has exceeded this level, the 180% ceiling is the *de facto* "market dominance threshold"; the carrier's (carriers') revenue/variable cost ratio on the traffic in question must be higher than this before market dominance may be found to exist.

Moreover, a mere finding of market dominance does not itself prove the rates are unreasonably high; it merely subjects them to regulatory jurisdiction, affording the complaining shipper the opportunity to submit additional evidence in hopes of persuading the Commission that the rates are actually unreasonably high. In the event, this is a virtually impossible undertaking. The Staggers Act provides that rates of railroads lacking "adequate" revenues are to be viewed liberally for regulatory purposes; and all major U.S. railroads were, as of this writing, deemed under ICC-established standards not to have adequate revenues. The Commission has devised a series of tests for the rates of such railroads which, in effect, allow the carriers to increase market-dominant rates to more or less any level they like.

Regulatory jurisdiction over the rates of surface common carriers of other modes (motor and domestic water carriers and freight forwarders) is not subject to the market-dominance limitation; they are, nominally, all subject to regulation.* Again, however, ICC policies have made it all but impossible to sustain a complaint that the rates of these carriers are unreasonably high; the presumption is always that the rates, being established in a market-competitive environment, are *ipso facto* at a "reasonable" level within the meaning of that term as employed in the Interstate Commerce Act.

Allegations of rate discrimination are equally difficult to sustain. In order for discrimination to exist, under current law and regulatory standards, conditions for the movements said to be discriminated against and the movements being compared must be absolutely identical—the same carrier(s), more or less the same route(s), the same traffic, etc., etc. Moreover, the ICC has adopted a "no harm, no foul" standard toward rate discrimination; the complaining shipper must not only prove the existence of discrimination but must also present affirmative legal proof that it has been economically injured thereby—an exceedingly problematic task.

Finally, it should be noted that, like shippers, competing carriers have been bereft of regulatory protections as a result of deregulation. In the regulation-constrained past, a carrier faced with what it considered "unfair" rate competition could complain to the governing regulatory body and seek an order requiring the complained-against competitor to raise its rates. Today this is no

---

*This does not apply, of course, to service exempt from regulatory jurisdiction, such as motor carrier transportation of agricultural commodities, etc. (*see Chapter 2*). Furthermore, once again the words "as of the time this was written" must be considered implicit in all discussion of non-railroad regulation, since it appeared likely, as this text was being finalized for publication, that there would soon be further, more-or-less total deregulation of all motor, domestic water and forwarder service.

longer possible. With respect to rail service, the "any rate" freedom is all-encompassing; it's obvious that the market-dominance exception can't apply in such circumstances (since if the complainant carrier is a competitor for the traffic there is *a priori* no market dominance). As to motor and domestic water carriage and freight forwarding, although regulatory jurisdiction nominally remains—in fact, these modes are expressly barred by the statute from establishing rates that will be "non-compensatory" (that is, will not return the costs of the service) or constitute "destructive competition"—in the first 5½ years after enactment of the 1980 Motor Carrier Act the ICC had not deemed even a single rate of any of these carriers to be unreasonably low.

(It should be noted that—once more, as of this writing—the situation was not necessarily as described above as regards intrastate motor carriage. In some states there was no economic regulation at all; in others regulation was essentially similar to what has been discussed; and in still others the regulatory hand limited carriers' and shippers' negotiating freedom far more stringently than for interstate carriage.)

With respect to airline rates, of course, there are no regulatory controls or limits whatever.

In sum, regulatory protections are, *de facto,* largely nonexistent for those who do not like the results of the marketplace in transportation. If shippers and/or carriers fail to negotiate satisfactory arrangements in the competitive marketplace, the government will not normally intervene to help them. This places, quite obviously, a heavy premium on negotiating skills.

Rate Bureaus

Before commencing any discussion of the actual negotiating process, one final question need be dealt with: In what setting should it take place? Outside the transportation industry this question is answered almost automatically; non-transportation business negotiations are conducted privately between (or among) the parties directly involved. In transportation, however, there is for surface modes an alternative that has a considerable tradition behind it: rate bureaus.

Bureaus have a long, and somewhat controversial, history in the transportation industry. They came into being during the industry's early years to support the practice of "collective rate-making," by which carriers created commodity classifications, service rules and standards, even actual prices on a cooperative rather than a competitive basis. By the 1940's bureau-based ratemaking was exerting a dominating influence over most common-carrier pricing.

Quite clearly this amounted to blatant price-fixing, and was successfully challenged in court as such by the Justice Department. However, with the backing of both carriers *and* shippers (who saw bureau ratemaking as a key stabilizing influence), Congress in 1948 passed the so-called Reed-Bulwinkle Act legalizing bureau activities (by granting them immunity from antitrust laws), by such a large margin that it became law over then-President Truman's veto. It was not until the mid-1970's that Congress reconsidered its stand, a process that culminated in provisions of the Staggers Rail Act and the Motor Carrier Act of 1980 that largely did away with the bureaus' ratemaking powers.

Prior to 1980 a substantial proportion of shipper negotiations with surface carriers—railroads, truckers, domestic water carriers and domestic surface freight forwarders—took place under the aegis of bureaus. Most such carriers engaged heavily (some almost exclusively) in bureau ratemaking processes, and the bureaus, both by ICC mandate and out of self-interest, encouraged shipper participation through their channels.

The 1980 laws and implementing regulations of the ICC—as well as a spate of antitrust litigation (*see Chapter 10*)—have done much to discourage both from working through bureaus. In the post-1980 era there has been a great deal more independent ratemaking by carriers (even though many carriers continue to have their rates published in tariff form by bureaus, and are encouraged by the ICC to continue doing so). And in these circumstances there is no perceptible advantage, and there are several disadvantages, to negotiating in a bureau setting.

First, contemporary bureau procedures (again dictated by

the ICC) no longer permit any carriers except those actively participating in the movement of particular traffic to take part in the negotiation of rates on that traffic. Indeed, the remaining statutory provisions that permit bureaus to exist at all as ratemaking entities do not even extend to open *discussions* of rate or service matters in which the discussing parties do not have a direct interest (such discussions being deemed in violation of antitrust law). Efforts by either carriers or shippers to gain input from non-participants as an aid to the negotiating process are thus fraught with peril under antitrust law and are likely, for that reason, to prove useless or worse.

Effectively, in sum, carriers must set their own rates without either consultation with or approval of their peers/competitors. This is so normal outside the transportation industry as to go without saying; but in light of long existence, and predominant ratemaking role, of rate bureaus, it is something that *does* need to be said, and emphasized, when it comes to this industry.

Second, negotiations in a rate-bureau forum must, under inflexible regulatory standards, be conducted in full public view—perhaps more than either party may like. Statutory limitations on the activities of bureaus (to the extent they are allowed to continue in existence at all) affirmatively *require* that the activities of these organizations be open to anyone who wants to know about them. With the eyes of potential or actual competitors sharply upon them, neither shipper nor carrier will be as able to negotiate as freely as if their talks were being conducted in strict privacy.

Third, such antitrust immunity as remains for the bureaus affords neither party any significant advantage. In some cases carriers will argue that this immunity is important where joint negotiations—those, that is, that involve interline service by two or more participating carriers—are being carried on. In fact, however, antitrust law nowhere bars *non*-competing suppliers from banding together to negotiate collectively with their mutual customers; thus, no such immunity is required.

And fourth, bureau activities are being subjected to extraordinary scrutiny by governmental antitrust enforcers and others. The opportunity for a misstep is too great in a bureau setting to warrant the risks of antitrust prosecution involved, for either party but particularly for the carrier (who, as vendor, is most vulnerable to antitrust law).

Thus, it will almost invariably be preferable for carriers and shippers to negotiate privately, one-on-one, than to seek to do so through the facilities of a rate bureau.

This should *not* however, be construed as discouraging or barring use of bureau rates as a basis for either the negotiations or even a final agreement. Especially with respect to smaller shipments moving by highway (less-than-truckload traffic), bureaus maintain rate levels, service parameters, etc., that are frequently used to fix *de facto* standards within the industry. That is, a particular carrier's tariff will frequently be based on nothing more innovative or individualistic than a specified discount off the bureau-published rate, perhaps with a few auxiliary service criteria. And bureaus remain entirely free to act as publishing agents for individual carriers' tariff provisions, either as separate documents or by means of incorporation in bureau tariffs themselves.

With this exception, however, both parties are best advised to eschew bureau participation; and the bureaus themselves should be regarded as no longer of significant consequence in the transportation ratemaking process.

4

MARKETING STRATEGIES FOR CARRIERS

"Marketing. . . is the whole business seen from the point of view of the final result—that is, from the customer's point of view."

So says famed managerial consultant Peter F. Drucker*, expressing an attitude which has become the starting point for today's sophisticated business marketing management. Modern companies dedicate a major part of their managerial resources to marketing in all its aspects—and, to the extent that they do so effectively, profit thereby.

From a marketing standpoint, however, the transportation industry is an isolated backwater. Decades of regulatory protectionism have allowed transportation carriers to neglect all but the most rudimentary marketing techniques and approaches without suffering economic injury. And even in the rare instances where carrier managers attempted to take an innovative (for transportation) approach to marketing, they were frustrated by the constricting coils of an python-like regulatory structure.

Consider what happened, for just one illustration, when the Southern Railway attempted to start up service to grain shippers in the early 1960's using its then-new "Big John" hopper cars. These cars, much larger and more capacious than previously available equipment, offered rail carriers a golden opportunity to reduce their operating costs and pass those savings (to at least

*Quoted from his *Management: Tasks, Responsibilities, Practices.* New York: Harper & Rowe Co., 1973.

some degree) along to shippers in the form of lower rates, thereby attracting additional revenue-producing traffic.

The reduced rates, however, first had to pass muster at the Interstate Commerce Commission, where they were, not surprisingly, vigorously opposed by competing carriers who foresaw the prospect of severe traffic losses. Even some shippers—those, that is, who had no prospect of benefiting from the reduced rates—joined in the fray, concerned that lowered transportation costs might afford their competitors a marketplace advantage.

It took *four years* to complete the ensuing litigation, during which the Southern saw its competitive lead erode daily. The first railroad unit-train proposal suffered similar problems and delays before the rates were permitted to take effect; so did the earliest grain mileage rates (dispensing with the now-forgotten "McGraham formula" rate-equalization complexities). Certainly such situations, which were repeated again and again, gave transportation carriers little incentive to be marketing innovators.

In these cases, at least, the innovating carriers did ultimately succeed in what they were trying to do; in others, even that consolation was denied. A decade after the "Big John" case two motor carriers devised the then-novel notion of offering shippers guaranteed delivery service; if shipments did not arrive at destination by a promised date and/or time, freight charges would be discounted accordingly.

The ever-meddlesome ICC would have no part of this. The law, it said, required all carriers to make delivery with "reasonable dispatch." And it read this as meaning that no carrier could seek to better the "reasonable dispatch" standard on its own—that, in other words, service guarantees were prohibited by law.

There is no need to belabor the many devices and rationales employed by regulatory authorities to discourage marketing innovation by transportation carriers. Those who were active in the industry during that period will recall them well; those who were not can only find this now-bygone era historically interesting, if indeed they are interested at all. Suffice it to say the net result was that carriers were aware from the outset that they would face major obstacles if they attempted anything in even the slightest way "new" to the industry—even if they sought to decrease rates below the level of their competitors!—and had, accordingly, a considerable incentive not to spend their efforts this way.

In addition, unlike other businesses, transportation carriers had the luxury of being able to discuss their rate and service levels freely with one another. By participating in rate bureaus they could escape the rigors of antitrust law and practice open collusion and "price-fixing."

To be sure, there was no legal or regulatory *requirement* that they do so; the freedom for carriers to set rates and establish service levels individually was always protected (subject, of course, to the oversight of the omnipresent regulatory structure). And many carriers took advantage of this opportunity—albeit generally only the smaller, less competitively important ones to any significant degree. But the lack of competitive pressure, again, gave carriers little reason to strike out boldly on their own in these areas; far safer, and generally more profitable, to merely follow the crowd and accept a parceled-out share of the competitively restricted market.

Finally, and perhaps most destructively of all to development of improved marketing techniques and practices in transportation, the regulatory structure fomented, even compelled, open antagonism between transportation carriers and the shippers who were their customers. The law reserved to the ICC (and, at the state level, the various state regulatory bodies) the sole prerogative to resolve problems and disputes which, in other industries, had to be settled by accommodation. And ICC proceedings were dealt with by means of the adversary structure of the law, under which the parties were obliged to engage in the litigatory equivalent of mortal combat with one another.

The result was that shippers and carriers, interdependent though they were, came most frequently to regard one another with dissatisfaction, distaste and disdain. The typical shipper-carrier relationship was like a bad marriage, in which the partners remain together out of custom or necessity, but leave no doubt, in the minds of one another or anyone else who happens by, how intensely they dislike both the situation and each other. Virtually any problem began in mutual acrimony and was settled (if at all) in open hostility.

In such a milieu it would have been ludicrous to suggest that carriers seek to view their businesses, as Peter Drucker urges, through the eyes of their customers. To a very real degree their customers were, as they saw it, their sworn enemies; and the lack of competitive pressure and opportunity gave them no encouragement to run up the white flag.

Admittedly some of the foregoing is exaggerated. Not every carrier, and not every shipper, partook of this attitude; and the experiences discussed above indicate that, although suppressed, marketing innovation was not totally absent from the industry. But it is with this background that one must view the revolutionary changes that have overtaken the transportation industry beginning in about the mid-1970's, and culminating in the major deregulatory laws enacted in 1978 and afterwards. Suddenly carriers have found themselves exposed to the exigencies of a competitive environment with which, in marketing terms, their past experience has left them ill-equipped to cope.

A Need for Rational Planning

Under deregulation, government has basically abandoned its long-asserted prerogative of restricting the carriers' marketing decisions; carriers today have virtually unconstrained freedom to make those decisions themselves.

Freedom is a word, and a concept, much revered in American society; paeans to freedom spangle our literature. But freedom has two faces; freedom to achieve is also freedom to fail. When deregulation removed the limitations on the carriers' managerial freedom to improve their marketing approaches, it likewise left them free to make marketing mistakes.

The key to successful exercise of the carriers' new marketing freedom is (as in almost any area of endeavor) clear, goal-oriented strategic planning. The carrier must seek to define, in as explicit terms as possible, what he wants out of any relationships into which he enters with shippers—how, that is, he seeks to be perceived in the marketplace. No longer does regulation ensure (to paraphrase a familiar saw) "a place for every carrier, and every carrier in its place"; now it is up to each carrier to define for himself that "place" he seeks to occupy in the market, and to defend his chosen niche against encroachment in a far more competitive environment.

Too often, especially in the early days of deregulation, carriers simplistically set their objective as maximization of revenues without considering the operational and other consequences of pursuing those revenues. Too many carriers grabbed eagerly and indiscriminately at the brass ring of every revenue opportunity that came their way, and only later (if even then) realized that such an approach can prove disastrous.

The classic illustration is, of course, Braniff Airlines. A great many factors underlay Braniff's decline into bankruptcy, but one of the most important was Braniff's overzealous pursuit of new markets. Excited by the opportunities deregulation afforded them, its managers set out to vastly enlarge the carrier's scope of service without adequately considering either the problems or the resource commitments such a course of action entailed.

Thus, Braniff jumped into market after market almost helter-skelter, with no clearly defined marketing strategy other than to expand as rapidly as possible. For the most part its decisions were tactically sound; that is, it was able to single out market areas where it could attract substantial traffic and, thus, revenues. But its scheduling, as it tried to make its fleet stretch among its growing number of routes and to avoid excessive "deadheading" or down-time in serving those often far-flung routes, became nightmarish. And its finances, as it sought ever more capital to support its expansion of its fleet and other facilities and to cover operating costs, grew increasingly precarious.

Ultimately the tightrope broke under it. Its overweening expansionism outreached its ability to adequately serve all the markets in which it was active; not only did it fail to achieve solid penetration into new market segments, but in trying to do so it neglected its old, established ones and found itself more and more competitively pressed there as well. At the same time, the severe economic recession of 1981-83 eroded airline traffic generally, hurting worst those carriers that lacked an adequate financial cushion—a category into which Braniff by now emphatically fell. The collapse was, in retrospect, inevitable.

Braniff's history emphasizes the need for carriers to develop clear-cut marketing strategies *before* they begin their search for those so-desirable revenues. No carrier can hope to be all things to all shippers, to provide all types of transportation service in all geographic areas. Each must, rather, find its niche in the marketplace and concentrate its energies and resources there.

This isn't to say, of course, that every carrier must narrow his operations to just a single type of service in just a single region of the country; certainly many carriers will do excellently with a considerable breadth of service and/or market areas. Just what market niche a given carrier can effectively serve—in terms of both type and magnitude—will depend on such things as:

- His available resources (financial, personnel, fleet, facilities, management, etc.). Quite obviously, no carrier should extend himself so far, or spread himself so thin, that he's unable to effectively meet his commitments while leaving a reasonable cushion to allow for the unexpected.
- His expertise and experience in different market areas. A carrier with no background in bulk transportation service should obviously think twice before invading this highly specialized market, for example; a local pickup/delivery trucker should give careful consideration before branching out into long-haul operations; and so forth.
- His ability to market his services effectively in his chosen area(s) of concentration. In the real world of business it's generally not enough merely to build a better mousetrap; one must also identify potential customers and persuade them of its worth. Invading a brand-new market area presents a major challenge to any company's sales/marketing staff, who must establish a level of trust and acceptance among an entirely new customer base.
- The calibre of competition he faces. It's obviously a lot

easier to make headway in a market where the competition is weak than in one where it is strong—especially if those strong competitors are poised and able to actively resist incursions. This doesn't mean a carrier should always shy away from markets where there's strong competition; but he should certainly go into such markets expecting problems and challenges beyond the norm.

• How well each particular operational segment can be integrated with others. An obvious example would be the carrier who latches onto a potentially lucrative haul from A to B, but lacks adequate revenue backhauls and finds his profits eaten up by excessive deadheading. The objective, after all, is not for particular operations to show a profit, but for the company *as a whole* to earn one.

In sum, the carrier must analyze, first, himself and his own capabilities and resources, and second, the marketplace and the opportunities *it* affords, and develop an overall marketing strategy he intends to follow. In the event unexpected opportunities come along that lie outside the scope of that strategy, they may, of course, warrant consideration; but they should not be blindly accepted without integrating them into a clear revised strategy.

Quite obviously, such a planning process must involve the entire organization. In particular, this process is not properly the exclusive province of, and should not be left wholly to, the Marketing Department. As with other operating departments, Marketing's role is to implement the company's policies and plans in its area of specialization; but the development of those policies and plans is an organization-wide responsibility, which must be overseen and guided by top management.

"Don't let your mouth write checks the rest of you can't cover," goes a popular saying. One of the most common mistakes of business is to do just that, by allowing Marketing (the "mouth" of the business entity so far as its customers are concerned) to make rash promises that the organization as a whole can fulfill, if at all, only at the expense of profitability. Certainly a strong, aggressive sales and marketing effort is a valuable asset for any company—but only to the extent that such an effort remains within the parameters of clearly established guidelines set with *all* aspects of the business taken fully into consideration.

Pricing Strategies and Approaches

Probably the most visible post-deregulatory trend in transportation marketing has been the reduction of rates and charges, often to extreme degrees. In some of the most competitive markets rates plummeted 30%, 40%, even 50% or more in the first few years after deregulation.

A great deal of this price-cutting was fomented, of course, by large shippers, many of whom used their newfound market "muscle" to force rates ever downward (to often ruinous degrees for the carriers). Making it known that they were, basically, "price-only" customers—that is, that price was their primary or sole basis for demand elasticity as regards any given carrier—these shippers produced a level of price competition rarely seen in the past in transportation.

Such shippers, however, paid too little attention to history. Similar "price war" conditions had occasionally existed in transportation's past—in the motor carrier sector prior to 1935 (when it came under regulation), for example, or in isolated pockets of the industry even during the regulatory era. One illustration of the latter circumstance, involving oilfield haulers during the early 1950's, typifies the almost invariable outcome of such situations.

At that time there was, for a variety of reasons, a considerable overcapacity in this transportation subsector. Then, as in the 1980's, the carriers saw their primary competitive tool to be pricing; spurred on by shippers, they became embroiled in reckless price-cutting in the hope of preserving their position in the marketplace.

But as rates tumbled ever down, the carriers came under more and more pressure to cut costs in order to keep their operations even marginally profitable. As a result, service and reliability (among other things) declined in virtual lockstep with rates. It took but a short time before shippers, realizing the toll this price warfare was exacting on their ability to fulfill their transportation needs, sought an end to the problem. In tune with the tempo of the times, they acted through regulatory channels; at their instigation, the ICC ultimately imposed a "minimum rate order" on the traffic in question to prevent further rate-cutting and encourage the carriers to rededicate themselves to providing adequate and reliable transportation service.

Much the same result emerged from the rate wars of the early 1980's, except this time by marketplace rather than regulatory means. Shippers came quickly to realize that excessive concentration on price as a sole buying determinant would result in unacceptably low levels of service. At about the same time, carriers, too, began coming to their senses, realizing that none of them could really hope to "win" an unremitting price war; at best they might gain Pyrrhic victories, the cost of which would ultimately precipitate the "winners" and "losers" alike into financial distress.

By the middle of the decade an uneasy equilibrium had been reached. Generally speaking, rates remained at appreciably lower levels (with the effects of inflation taken into account) than before deregulation; and there remained vigorous price competition. But except for sporadic outbreaks of renewed warfare in the most competitive markets, price competition settled—as it has in other economic sectors—within a fairly narrow range. Carriers became far more cost-conscious in their pricing decisions; increasingly they began basing those decisions on their need for profitability rather than just blindly pursuing traffic and revenues without regard to the effect of that pursuit on the economic "bottom line."

In sum, rate-cutting was coming, as these words were written, to assume its realistic place in the carriers' arsenal of marketing weapons—an important approach to the marketplace, but by no means the only one. And there was also evidence that the carrier industry was beginning to segment itself into a variety of niches based on differences in the caliber of service and the level of rates.

The clearest example of this comes, again, from the airline

sector (not surprising, in view of the fact that airline deregulation predated deregulation of other modes by two years or more). There have evolved at least two (some would say more) different types of air carriers, one of which continues to regard pricing as its main marketing implement while the other focuses more heavily on service. The former group is typified by People Express, which offers few amenities and bare-bones service but charges extraordinarily low fares; the other is comprised mostly of older, established trunk lines, which for the most part don't attempt to compete with the bargain-basement prices of People Express and its brethren but base their marketing strategies instead on service-related factors—greater comfort, availability, reliability, etc.

Certainly there would appear to be room in the transportation industry—as there has proven to be in most other business sectors—for at least these two, and perhaps more, pricing/service strategies. But as in other economic sectors, it is necessary for any participant in the marketplace to clearly position itself; no one concern can reasonably expect to meet the needs of *both* those seeking a high quality of service (and willing to pay a higher price for it) *and* those whose primary emphasis, as buyers, is on price. The degree to which a carrier may expect to succeed in his marketing endeavors may depend in large part on how clearly he defines his role in the price/service spectrum, and how clearly he conveys the message of where he stands to the market he seeks to serve.

As price competition has become more important in the transportation industry, carriers have adopted different approaches to presenting their prices to the marketplace. The effort, of course, is to make prices *seem* lower, from the customer's perspective, than they really are, and thereby more attractive (at least so the carrier hopes) than the price offerings of the competition.

Unique to the airline industry (at least as of early 1986, when this was written) is the "restricted" fare—the heavily advertised rock-bottom fare that applies only under highly limited conditions. An airline may offer, for example, a fare that applies only if the reservation is made and the ticket purchased 30 days in advance of the travel date, subject to a substantial penalty for cancellation or change of travel plans, and limited to but a proportion of the available seats on any given flight. (To satisfy truth-in-advertising legal standards, such ads usually contain fine-print disclaimers about "subject to conditions" and "limited availability.")

Such "low-ball" restricted fares also serve a second purpose, one with little application outside the passenger sector of transportation. The airlines are seeking to attract new, discretionary business—to lure individuals who otherwise might not travel at all. Since freight transportation for the most part is not of a similarly discretionary nature—whether goods will or will not move generally is a function of factors other than transportation cost—this purpose would largely not be served by such restrictive rate practices by freight carriers, which is probably responsible for the absence of this type of ratemaking outside the passenger-focused airline sector.

An alternative approach which attained much favor among freight carriers (mostly in the trucking sector) in the early 1980's was discounting. Rather than express his rates on a single-factor basis, a carrier would establish a level of (relatively high) rates with one hand and then offer substantial discounts off those rates with the other. There are several marketplace considerations that support this approach to ratemaking:

First, from the perspective of many carriers this approach greatly simplifies the process of setting rates and publishing them in tariff form. Notwithstanding their dwindling, antitrust-restrained role as ratemakers (*see Chapters 3 and 10),* motor carrier rate bureaus continue to publish tariffs setting forth rate scales which their carrier members are free to adopt. Since deregulation, these scales have been maintained at, for the most part, an unrealistically high level, so that little, if any, traffic actually moves under them.

In order to establish equivalent, competitively lower scales of their own, however, carriers would have to undertake much work, and publish extremely lengthy individual tariffs. Rather than going to all this trouble, many truckers have chosen instead to adopt the bureau rates subject to extensive discounts—a practice which has the added advantage of not presenting shippers with the confusing and laborious task of researching each such individual carrier's tariff, but instead allowing them to check only competitive discounts against a single rate-scale publication.

Second, as retailers have long recognized in other economic markets, discounting often provides a psychological sales "edge." It has been proved again and again that an item advertised, for example, at "50% off!" a supposed $200 price will outsell, by an appreciable margin, the same item advertised at a simple $100, even though the customer pays the same in either case.

Industrial traffic managers are far from immune to this ploy. Indeed, where carriers have sought to get away from discount structuring of their rates and to instead offer reduced scales with discounts already "built in," they have suffered significant traffic losses; in some instances this has even happened when the revised non-discount scales actually produced *lower* charges than the pre-existing discounts. This seeming paradox owes much to the fact that many shippers measure their traffic departments' performance managerially based on the "savings" those departments can document, and such documentation is commonly based simply on discount levels rather than on actual rate comparisons.

Discounting also simplifies a carrier's ability to negotiate different levels of rates with different shippers without having to prepare voluminous tariffs or contract rate schedules for each. Thus, a carrier might accord one shipper a discount of 35% off bureau rates, a second a 40% discount, a third 30%, and so on, merely by adding a few brief lines to a tariff or contract; whereas to publish pre-discounted rates anew for each shipper would require many pages.

(Another approach to this customer-differentiated discounting is the "participatory" tariff, whereby a standard level of discounts is offered to all shippers but is only available to shippers who have expressed, in advance and in writing, their desire to "participate" therein. Quite obviously it is the regular user of the carrier's service who will be best positioned to take advantage of such "participatory" discounts; the occasional or one-time shipper may not even know of the existence of the discount potential, and may also not have enough advance notice of its shipping plans to meet the write-in requirement.)

Some shippers are also attracted by the fact that discounts are not always fully reflected on the freight bill or other shipment documentation. This happens when carriers adopt a variant form of discounting by which actual billing is based on undiscounted rates and charges, with the discount either accommodated through a separate billing adjustment before charges are paid or refunded separately after payment.

The reasons underlying this preference by shippers are occasionally somewhat unsavory. In many industrial buyer-seller relationships the party actually bearing the freight charges is not the party who negotiates with carriers. In these circumstances, the fact that discounts may not show up on the freight bill clearly affords the non-paying negotiating party an opportunity to, in effect, reap a surreptitious (and generally unearned) profit at the expense of the other *(see Chapter 5).*

As "unbilled" discounts grew increasingly prevalent in the trucking industry in the mid-1980's, it became evident that such practices were not uncommon. In some business sectors the situation had come sufficiently to light to generate considerable buyer-seller recriminations and disputes, to the point where threats of litigation over the hidden discounts were being bandied about. However, it appears that carriers themselves are immune from any possible legal repercussions; indeed, the ICC has expressly countenanced the legality of carriers' paying discount-based "refunds" to the negotiating rather than the paying party. Accordingly this type of discounting practice remained a part of many carriers' marketing strategies as of the time this was written.

An especially interesting feature of post-deregulatory price-cutting and discounting has been the heavy focus on line-haul rates, fares, etc., of the carriers. Other, ancillary charges—although they may in some cases account for a considerable proportion of the total freight bill—have been largely ignored.

In a few areas this has been accompanied by growing shipper pressures on carriers not to bill such ancillary charges at all. Motor carrier detention charges, for example (assessed where carrier equipment is held beyond allowed "free time" for loading and/or unloading), are frequently overlooked by unspoken agreement between shippers and carriers in all but the most extreme circumstances. (Railroads have been appreciably less amenable to such pressures in connection with either freight car demurrage or detention charges on TOFC trailers.) Sometimes this also extends to charges for such services as diversion, reconsignment, stopping in transit, etc.

A more important facet of the situation, however, has been the general focus of the shipper community on line-haul rates, to the exclusion of all other charges, in shopping for transportation service. As (again) retailers in other economic sectors have long recognized, where multi-factor pricing is involved it is a common phenomenon for buyers to pay more attention to the largest single component of a bill than to other, lesser components—or to the aggregate amount to be paid.

In some cases carriers have reacted by adopting the tactic of paring away "standard" ancillary services at the same time they reduce rates or fares. Again, the best-known examples come from the airlines, where first the now-defunct Laker Skytrain and then People Express began charging extra for such services as checked baggage, beverage and meal service, etc.—all provided at no additional cost by most other carriers—while cutting their "base" fares; indeed, in some cases airlines even charge more for confirmed reservations than for "walk-on" passengers. Among motor carriers of freight, some have eliminated such services as carrier loading and unloading and the like unless the shipper pays extra.

On the other hand, relatively few freight carriers, at least as of the time this was written, had adopted the obvious alternative of using increases in accessorial or other charges to help offset line-haul rate reductions. It appears that most carriers share the shippers' mentality in regarding line-haul rates as their main revenue source, with other services priced largely based on the cost of providing them rather than their revenue potential.

A particularly surprising aspect of this relates to credit terms. In most businesses management is acutely aware of the value of money, and reflects that in credit policies and terms; in transportation, however—probably due in large part to the industry's regulatory history—this awareness is perceptibly less acute.

Leaving aside the special case of air passenger service, regulatory authorities for decades imposed exceedingly, and unrealistically, strict limits on the extension of credit by transportation carriers. The rules provided that railroad and freight forwarder shippers must pay all freight charges within five days after bills were submitted; the customers of motor and domestic water carriers had but two additional days in which to pay up; and similar restrictions were placed on the customers of other regulated carriers.

But since the regulatory agencies had jurisdiction only over the carriers, and not their customers, the rules necessarily had a glaring loophole. The agencies could, and did, require that the carriers collect their money within the specified time limits; but they could not, and did not, require that shippers *pay* within those limits. There was thus no effective enforcement of the credit standards imposed by the regulators; and in light of the exceptionally brief credit periods allowed by the rules, they were widely ignored by all; it became not unusual for shippers to take 30 days, 45 days, 60 days or even longer to pay freight bills.*

It should be obvious that this situation could only have a

*Technically, carriers were obliged to enforce the rules on their customers by the expedient of cutting off the credit of any shipper that did not comply. However, any carrier who did so knew full well he would also be cutting off that shipper as a source of future freight (and revenues), and thus took such action only in the direst circumstances—about the same circumstances, that is, as would trigger a similar reaction in any other economic sector. Since only this "all-or-nothing" reaction was countenanced by the regulators—delinquencies could not be penalized, as they are in other businesses, by the more realistic and appropriate means of late-payment surcharges or forfeiture of prompt-payment discounts—the credit rules became *de facto* toothless.

highly negative effect on carriers' cash flow. Yet, even though all carriers now have the opportunity to take more effective action to limit credit to shippers (the ICC, for example, radically liberalized its credit rules in May, 1985), relatively few, as of this writing, have exercised their new freedom. The prompt-payment discount, the late-payment penalty—the traditional devices of other economic sectors to cope with this problem—remain rare in the transportation industry; nor will a shipper's payment policies, whether prompt or slow, normally be taken into account in a typical rate negotiation.

Self-evidently, such a state of affairs does not appear likely to continue indefinitely. But once more the transportation industry's regulatory history and traditions have made it extraordinarily slow to respond to economic pressures to which the rest of the business community, lacking that history and those traditions, has long addressed itself. Ultimately, as in other economic sectors, it appears inevitable that credit will become competitive and will play its role with other pricing factors in the marketing policies and strategies of the transportation carriers.

The Tactics of Marketing

Beyond "pure" pricing (and/or discounting), the first few years of deregulation have seen a remarkable array of marketing variations on the part of carriers. Some represent broad efforts to put marketing more in tune with customer needs, while others are of a more "gimmicky" nature. Unsurprisingly, virtually all have been adapted from the practices of other business sectors, which never felt the heavy hand of regulation to restrict their freedom to market their wares (whether goods or services) as they saw fit.

In no particular order, and with no pretension that this represents anything approximating an exhaustive list, here are some of the approaches that have been tried:

• *Volume ratemaking and/or discounting.* Always a factor in transportation, this has become much more important in the post-deregulatory era due to the increased market competition.

As it is elsewhere in the economy, security—the ability to rely on a stable base of revenues—is of great importance to carriers, and something for which they are willing to pay in the form of reduced or discounted rates. Many carriers even offer "tiered" scales of volume-based rates—one level based on very low periodic volume, a second (lower) one for a little higher volume, a third (still lower) for more volume yet, and so on.

• *Percentage-of-volume rates.* A variant on volume ratemaking, this involves conditioning the shipper's entitlement to volume-based rates or discounts on a "floating," rather than fixed, volume of traffic. That is, the shipper undertakes to give the carrier a set (usually high) percentage of its total traffic over a particular route or routes, rather than any absolute amount of traffic. From the carrier's perspective this affords less security than does the fixed-tonnage volume standard (since if the shipper suffers unforeseen setbacks in its own market there will be less traffic in which the carrier will share); but the other side of the coin is that the carrier may also reap unexpectedly high traffic, and revenues, if the shipper's business (and therefore its traffic) increases unexpectedly. In effect, carrier and shipper become, in a very real sense, partners in both the risks and the rewards of the shipper's market.

• *The "frequent flier" program.* This tactic is identified in accordance with the terminology used by passenger airlines, who most commonly employ it; but variants on the same theme have been adopted by other, freight-oriented modes as well. In such programs various, mostly "in-kind" awards (usually free or reduced-price tickets, in the case of the air carriers) are offered to customers who make frequent use of a particular carrier's services. The airline awards have caused much consternation because awards are always given to the individuals who actually travel, even though, in many instances, it is the travellers' employers who are actually paying the bills.* Among freight carriers, some

*Many companies, as well as (in its capacity as employer) the U.S. government, purport to demand that their employees turn over to them frequent-flier awards deriving from business-related travel, rather than retaining the awards

have followed the airlines' lead literally, offering awards to individual industrial traffic managers (rather than their shipper/employers), which has led to the same problem; in most cases, however, carriers give awards under such programs to their corporate shippers rather than those shippers' employees.

• *Lotteries.* Not dissimilar to the frequent-flier-type programs is the notion of lottery-based awards pioneered in transportation (shortly after deregulation) by North American Van Lines. During a specified period of time, in NAVL's version of this concept, the names of all shippers of household goods via that carrier were dropped into a hopper; at the end of that span one name was drawn at random. The happy winner received a refund equal to the freight bill he had paid for his move (or, if he hadn't paid yet, a waiver of the bill). Similar lottery-type programs have since been instituted by other freight carriers; in most cases the award has been a waiver or reduction of freight charges, but in a few it has been in the form of prizes given to individual shipper employees, leading to the same sort of disputes as have arisen under the frequent-flier-type programs.

• *Guaranteed delivery service.* Rejected out of hand by the ICC during the era of strict regulation (*see above),* this approach has resurfaced under deregulation in a variety of guises. At its most extreme, carriers offer to waive 100% of their freight charges if delivery is not made within the specified time. More commonly, late delivery entitles the shipper to percentage discounts off the freight bill, which vary according to the degree of lateness.

• *Other service incentives/penalties.* Guaranteed-delivery service is not, of course, the only form of service-related incentive and/or penalty ratemaking in use. In particular, rail carriers sometimes offer incentive/penalty arrangements in connection with switching service and equipment availability, where ship-

for their (the employees') own use. In some instances, however, employees have refused to comply, which has led to both morale-destroying disputes and a great deal of employee deceit.

pers are concerned that they may not be able to get equipment adequate to handle their traffic when they want it. Airlines' practices of offering un-accommodated passengers monetary compensation for their inconvenience—originally mandated by the CAB, but now maintained voluntarily by the carriers, for marketing reasons, after the regulatory agency has been abolished—are another illustration of this approach.

• *The "special introductory offer."* Typical was the approach of Roadway Express, one of the nation's largest motor carriers, when in 1983 it sought to establish itself in the Pacific Northwest area—an area it had previously served but little. It declared a "grand opening" of its service in that area, and advertised widely that, for the first 60 days following that date, it would provide such service at half price—50% off its otherwise applicable rates and charges. The idea, of course, was to break the allegiance of shippers to the carriers that had previously served them, in the hope that, once the 60-day introductory period had expired and the half-off discount was withdrawn, Roadway would retain a significant proportion of the freight. A number of other carriers of the same and other modes—especially airlines seeking passenger traffic over newly opened routes—have taken similar actions.

• *The "loss leader."* This is a form of the cross-subsidization long decried in transportation, when it was enforced by regulatory agencies who insisted that carriers employ the profits from higher-margin market segments to finance unprofitably low rates elsewhere. The regulators made such decisions for "social" reasons, seeking to mastermind what they deemed "best" for the economy as a whole; but now that their heavy hand has been lifted, some carriers are voluntarily adopting the cross-subsidization principle on other, market-related bases. Typically, a carrier will agree to handle certain of a shipper's traffic at a low profit margin—perhaps even at a loss—in exchange for a commitment that he will also be given other, more profitable traffic to haul. An occasionally used variation involves providing low-profit or unprofitable service today in exchange for the hope that it will generate more profitable traffic later on—as where a carrier reduces rates temporarily to allow a customer to gain a footing in a new geographic market area of its own, subject to the shipper's commitment that, once the new market is "cracked," traffic to it will continue to move via the same carrier at, now, higher and more profitable rate levels.

• *"Traffic-balance" ratemaking.* One-way traffic lanes are the bane of any carrier's existence; he has a goodly volume of potentially high-profit traffic moving from point A to point B, but little or nothing on return, so that his equipment, operating personnel, etc., are productive only half the time. In some instances this will result from factors affecting only a given carrier; in others it will be the product of regional peculiarities (for example, south Florida is notoriously a "consumer" region, etc.). Either way, carriers will often reduce rates markedly in the low-volume direction to improve their round-trip balances and avoid costly backhaul "deadheading" (or, alternatively, having to give up some or all of their lucrative fronthauls).

• *"Off-peak" and "seasonal" rates.* These time-oriented versions of traffic-balance ratemaking are especially prevalent among passenger airlines and household goods movers. The former experience multi-faceted variations in demand based on time (weekend and holiday-oriented travel peaks, unpopularity of late-night flights, etc.), while the latter have huge seasonal differentials (the May-September period, when school is out, is the traditional time for moving); and both seek to compensate by offering price-based incentives for their customers to choose less popular times. Similar incentives are, more and more, being offered by other carriers serving markets that exhibit distinct seasonal or peak/off-peak patterns.

• *Rate "averaging."* This has taken a variety of forms. Geographically speaking, the most obvious illustrations are the motor carrier ZIP rates, where all points within a given area (defined by the Postal Service's three-digit ZIP Codes) are accorded the same rate treatment, even though the area may encompass tens of thousands of square miles. With respect to commodities, unrestricted (or lightly restricted) freight-all-kinds (FAK) rates are being offered by carriers of all modes with increasing frequency. First Clipper Exxpress and subsequently other freight forwarders have taken an even broader approach to the concept of averaging, combining both geography and commodity factors and including the additional element of weight; for some shippers they simply average out charges on all shipments for a past period of time and then apply that average as a flat charge on future shipments. In any form, of course, averaging greatly simplifies the often extremely complicated bases on which transportation services have been traditionally priced—a particular advantage to carriers, since (as vendors in other economic sectors have long recognized) a simple pricing policy is generally more attractive to customers than a complex one.

And there are, of course, many other types of innovative marketing tactics being employed by carriers.

The Special Case of Loss-and-Damage Liability

One additional option increasingly offered by carriers concerns the extent to which the carrier will be liable for loss of, damage to, or delay of the goods while in the carrier's possession.

Both the common law and the Interstate Commerce Act provide for more or less absolute carrier liability, irrespective of the cause of loss, damage or delay. The carrier serves as "virtual insuror" of the goods, excusable only where the loss/damage/delay resulted from (1) an act of God; (2) act of a public enemy; (3) act of a public authority; (4) act or omission of the shipper (or consignee); or (5) "inherent vice" (intrinsic nature or characteristics) of the goods themselves. If he cannot prove one of these causes was 100% responsible—and, in addition, cannot submit proof that he was himself not negligent in his handling of the goods—he is, by law, liable for the full amount of any economic injury suffered by the goods' owner.

For decades, however, carriers have been permitted to limit the monetary amount of their liability provided that, in exchange, they offered shippers the benefit of reduced rates and, additionally, obtained the shippers' written agreement to accept such reduced levels of liability. But these "released rates" required, for regulated carriers, the approval of the governing regulatory body—approval that was sparsely granted.

Today, deregulation has made regulatory approval of released rates unnecessary—and has also afforded carriers the opportunity to reduce their liability in other ways. As a result, the number of released rates in existence has mushroomed in the post-deregulation era; and so have the varieties of liability limitations. Some carriers now, for example, disclaim loss-and-damage liability except where they were negligent—and lay on shippers the obligation to prove negligence. Others restrict time limits for the filing of loss-and-damage claims, or establish "deductibles" below which they will not entertain loss-and-damage claims, or impose other forms of liability limitations.

All such limitations are, of course, subject to the requirements that (1) the shipper receive, in exchange, some offsetting benefit, usually a reduced rate, and (2) the shipper agree in writing to accept the lower liability limit; but in some instances carriers' alternative, full-liability rates are so prohibitively high as to leave shippers with little realistic alternative but to accept the reduced level of liability.

This is seriously disturbing to many shippers. They are accustomed to the common-law standard of carrier liability—which, after all, has been embedded in the law for centuries predating the existence of the modern transportation industry—and feel they are being deprived of their legal rights if carriers do not live up to that standard. Moreover, some shippers worry that, if carrier liability standards are reduced, the carriers' incentive to care for the goods they are transporting will be likewise reduced, resulting in appreciably higher levels of loss, damage and delay.

Such prejudices, however, ignore economic reality. The question shippers should be asking themselves is whether they are better off purchasing insurance from a transportation carrier not in the business of assessing and placing an economic value on risks of loss, damage or delay. . . or whether they should instead depend on insurance underwriters who *are* in this business. The shipper who naively believes he is not, one way or another, paying his transportation carrier(s) for assuming full carrier liability—usually by means of an increment built into the transportation rate level, but sometimes separately stated in the freight bill—is simply deluding himself. He is indeed paying; and in most instances the premium he is paying is a great deal more than would be charged by commercial insurors, who have both a better ability to assess and value such risks and a greater financial "base" over which to amortize them.

Carriers for the most part tend to identify different traffic groupings (defined in different ways, carrier by carrier) as independent "profit centers" for managerial and accounting purposes. Thus, the risk of loss-and-damage claims—including possible "catastrophic" claims of perhaps hundreds of thousands, or even millions, of dollars—must be amortized, insofar as is possible, within the framework of only a portion of the carrier's aggregate revenue potential. Accordingly, because the "base" is relatively small, risk-assessment involves placing a fairly high premium on L&D liability from the carrier's managerial perspective.

To illustrate with a situation drawn from litigatory annals: When the Washington, DC, subway system was being constructed, the passenger cars that would be used on it were built in Atlanta, GA, and had to be transported from there to Washington. The cars were not suitable for long-haul rail movement on their own wheels (they were not built to withstand the physical stresses of ordinary rail operations), and hence had to be loaded aboard flatcar equipment for their Atlanta-to-Washington moves.

The subway cars were of considerable value—well over $100,000 each. The rail carrier (one of the nation's largest) that provided the transportation service offered the manufacturer/shipper an extremely favorable rate, *provided* the shipper agreed to limit the carrier's liability in the event a derailment or other untoward event resulted in a catastrophic claim. Although the shipper agreed, the ICC—which at that time (mid-1970's) still had regulatory jurisdiction over such matters—refused to ratify the agreement, and the carrier was obliged to transport the subway cars under full carrier liability. Fearing the impact of a possible major claim on this "profit center" series of movements, it would do so only at rates approximately *double* what it had offered if the shipper agreed to limited liability.

The carrier's position in this case is certainly understandable, considering how it regarded the movements in question (indeed, the ICC declared its full-liability rates to have been reasonable, from a standpoint of regulatory law, when the shipper later challenged them in litigation). Nevertheless, the cited example demonstrates how L&D liability can distort carrier ratemaking.

If even a major carrier (to whom the most major subway-car claim would have been financially inconsequential in terms of the carrier's overall financial picture) is so fearful of the effects of a catastrophic claim, it follows that reduction or elimination of the extraordinary liability the common law otherwise (in the absence of "released rate" agreements) imposes on carriers is, from their standpoint, greatly to be desired. Now that the law increasingly permits such agreements, thus, it is scant wonder that a growing number of carriers are actively promoting this alternative in their marketing of their services to shippers under terms that are mutually beneficial to both parties.

And from the shipper's point of view, the idea that carriers will take less care of goods if they do not have full liability is not borne out in either theory or practice. Quite apparently, shippers are hurt by loss of, damage to or delay of their goods even if they are, ultimately, fully compensated for their out-of-pocket losses. They, or their customers, do not have the goods in a timely manner, and cannot therefore use them for the purposes for which they were intended. A poor loss/damage/delay record or reputation will thus normally be a strong competitive strike against a carrier; and, in an unregulated marketplace-oriented environment such as the modern transportation industry now constitutes, carriers in such a position cannot expect to prosper.

Even such an authority as the Shippers National Freight Claim Council—whose policy is one of unrelenting opposition to limitations on carrier liability—acknowledges that carriers' performance in this area did not deteriorate in at least the first few years of the post-deregulation era. Notwithstanding the fact that released rates, in their various forms, have multiplied vastly following deregulation, the SNFCC admitted in 1985 that loss, damage and delay claims actually declined over the same period.

None of this is to suggest that shippers should automatically accept released rates where offered the requisite choice. It does, however, indicate that where the option is offered it should be given serious consideration, and the ultimate decision should be based on economic considerations. Is it cheaper to pay for full carrier liability? Or will the shipper gain an economic benefit by accepting limited carrier liability and covering himself by means of his own separately purchased insurance (or, if he chooses, self-insurance)? Sound negotiating technique demands that this question, just as others associated with the negotiations process, be evaluated by both parties from a strictly economic point of view.*

The Art of Transportation Salesmanship

The sales representative is the "point man*" of the carrier's marketing program. It is he who represents the carrier in both its first and its continuing contacts with its customers—he who, in a very real sense, *is* the carrier in the eyes of the shippers who do business with it.

---

*It should be noted that this approach to questions of carrier liability and insurance has long been the *de facto* standard of the maritime sector of transportation, where laws and international treaties afford so many loopholes through which carriers may escape loss-and-damage liability that most shippers routinely purchase separate "all-risk" coverage for their goods while in transit. This is not to suggest that the same approach is *ipso facto* desirable for shippers by and carriers of other modes; but it does indicate that blind adherence to the strictest standards of carrier liability is not necessarily a prerequsite for either carrier marketing or shipper purchasing policies.

*Once again, the masculine gender is employed in this text merely as a literary convenience to avoid such awkward locutions as "salespersonship," "point person," etc., and is intended to imply no prejudices as to the sex of actual persons working in the transportation industry.

Yet salesmanship has been one of the most neglected areas of the carrier industry's management.

Typically, shippers complain that carrier sales personnel are unknowledgeable about details of the service their employers offer, and are ill-equipped to deal with any special transportation needs they (shippers) may have. This should not be construed as a slur on the individuals who occupy these roles; rather, it reflects instead on carrier management practices and predilections.

In the ordinary organizational structure of transportation carriers, sales representatives have only a secondary (at best) role. Most carriers view the selling of their services as largely unrelated to what they deem their primary business of *providing* those services, and in consequence downgrade this function severely. Sales representatives are afforded little, if any, training in the details of the carrier's operations; their primary source of information about the carrier is the same promotional/advertising material they, in turn, pass on to shippers. And they usually receive scant attention when they pass back information gleaned from customers which, if heeded, might help to fine-tune the carrier's marketing strategy in important ways.

That sales should be thus shunted one side within the managerial structures of carriers is not altogether surprising. The apocryphal "travelling salesman" jokes and other popular disparagements attest to the scorn with which the selling profession is traditionally regarded. And the regulation-induced marketing laxities of the transportation industry have given carriers little incentive to move, in terms of managerial attitudes, beyond this hoary tradition.

Modern managerial theory recognizes, however, that the sales representative has (or should have) a far more important role than merely as the butt of stale humor. It is, after all, he, more than any other individual, who shapes the way an organization's customers perceive that organization; and accordingly he is, and should be viewed as, the focal point for its tactical marketing program.

It is not the place of this text to offer an extensive dissertation on the techniques of salesmanship; that topic is addressed by innumerable treatises devoted exclusively to it, which have as much application in transportation as in other economic sectors. There are, however, certain common-sense guidelines that may be identified:

First, the carrier sales representative should be well schooled in both the transportation industry in general and the operational particulars of his employer. Where he encounters shipper personnel who lack such knowledge, this will enable him to help them articulate their own transportation needs and, in the process, to acquire traffic for the carrier he represents; where he meets industrial traffic managers with expertise of their own, it lets him do so on more level terms and respond usefully to their specific questions and needs. Additionally, sales training programs should expressly include the skills and techniques of business negotiation, an area in which promotion-oriented sales personnel are traditionally weak.

Second, he should be supported by solid research concerning the shippers on whom he calls—and should be encouraged to augment this informational base with his own knowledge and ideas. If the Socratic "know thyself" is the first rule of salesmanship in any area of business, surely "know the other party" must be deemed the second. The salesperson who walks into a customer's or prospective customer's office "cold" is at an all-but-insuperable disadvantage *vis-à-vis* his competitor who can talk intelligently with the customer about his (the customer's) own business, needs, wants, etc.

Third, the sales representative is the individual "on the scene" in the marketplace; information he brings back from his forays should be attended carefully. Especially should management listen when he relays customer problems or complaints. The salesperson to whom a customer is assigned is likely to have far greater rapport with that customer than anyone else in the organization, and can often be extremely valuable as an "early warning system" for matters that might, if left untended, ultimately impair or destroy the relationship altogether.

(This emphatically does *not* mean, however, that management should always follow its sales representatives' recommendations unhesitatingly. Sales personnel characteristically see their customers from an exaggerated perspective, out of both psychological identification with those customers' viewpoints and a fear that they might lose the customers and, consequently, suffer a cut in their personal income. The salesperson who panic-strickenly reports that he's "about to lose this shipper" unless major concessions are made may be telling the truth—and then again he may be vastly overreacting to what was merely a bit of grumbling on the customer's part (or, of course, anything in between). It is of course advisable to follow up on such situations; but care should be taken to ascertain the facts fully before precipitate actions, of any sort, are taken.)

Fourth, the sales representative should be invested with at least some degree of authority to negotiate with shipper personnel. Obviously, the salesperson who is perceived by customers as little more than a glorified errand-runner—competent only to carry messages from and to his principals—will have little standing in the eyes of those customers, and this attitude will inevitably reflect on his employer. In addition, a great many business deals and relationships have their roots in such *ad hoc* "spot" negotiations. To be sure, such authority cannot be limitless; the sales tail must not be allowed to wag the organizational dog. But within the parameters of established marketing policy sales representatives should be afforded a reasonable degree of flexibility to do business on behalf of their employers.

Fifth, where negotiations go beyond casual "spot" business dealing, the sales representative to whom the account in question is assigned should be accorded a prominent place at the bargaining table. He will not, of course, be the carrier's "lead" negotiator, since such negotiations by definition go beyond the scope of his personal authority; but he should be accorded some voice in the decision-making processes attendant to such negotiations, and, moreover, should be *perceived by the shipper* as a key member of the carrier's negotiating team. This will afford him much greater standing in the shipper's eyes, both in terms of helping to facilitate any agreement that may be reached and in garnering future, additional business from that shipper.

Much more, of course, could (and should) be said about transportation salesmanship; and readers are strongly encouraged to seek out more detailed information on this question. One additional point, however, warrants mention here because of the impact deregulation has had in this particular area.

For decades virtually all forms of business entertainment and gift-giving—commonplace in other economic sectors—were strictly prohibited in transportation; such practices were viewed as the granting of illegal rebates. Illustrating the rigid application of this regulatory policy was the case of a motor carrier that arranged a "junket" to the Kentucky Derby for traffic managers of several major shippers in its area of service (including, it should be noted, some for whom it had never transported any traffic at all).

In ensuing court proceedings instigated by the ICC's enforcement staff, it was held that the carrier had violated the anti-rebating Elkins Act. A particularly interesting feature of the court's ruling was that the violations concerned *both* shippers the carrier was then serving *and* those for whom it had never transported even a single pound of freight; as to the latter, the court deemed the excursion to constitute a before-the-fact rebate on putative *future* shipments the carrier might haul.

Deregulation eliminated these constraints on carrier "business entertainment," placing them on the same footing in this regard as vendors elsewhere in the economy. In some quarters this has tended to encourage an overreaction, with carrier sales representatives "going overboard" in entertaining present and prospective customers lavishly, giving them valuable gifts, etc.

Contemporary social attitudes frown on such practices. This is reflected in the policies of many industrial concerns, which prohibit their employees from accepting more than token gifts (calendars, paperweights and the like) from vendors or potential vendors; some even go so far as to ban the traditional business lunch, paid for by the vendor's representative. The same thinking also underlies the so-called "commercial bribery laws" of the various states—to which transportation shippers and carriers are subject just like buyers and sellers in other economic sectors—many of which have been tightened appreciably in recent years.

Moreover, there can be no question that the "fat" sales expense account can seriously erode corporate profits from the business supposedly gained by means of entertainment and gift-giving.

In most business relationships today such practices play little, if any, role in the sales endeavor. For all of the foregoing reasons, plus others that will readily come to mind, the carrier industry is well advised to follow this lead notwithstanding the regulatory relaxation in this area.

5

BUYING TRANSPORTATION SERVICE: THE SHIPPER'S VIEW

"If you don't know where you're going," runs the old saw, "any road will take you there."

A successful business negotiation, in transportation as anywhere else, depends far more on what the parties do *before* they sit down at the bargaining table than on what takes place at that table. In motion pictures, plays, books, etc., depicting the negotiation process (whether real or fictitious), the emphasis is always on clever tactics and on-the-spot decision-making during the actual negotiations; that, after all, is where the visible drama lies. Real-world businesspeople who try to negotiate in that fashion, however, will find that all this fine dramatic effect merely disguises a lack of substantive results; not having predetermined what their negotiating objectives are, they'll almost inevitably fail to achieve what they really want or need.

The negotiating process, thus, must start with "homework"—a clear, rational analysis of where the company wants to go, what its objectives really are—what, in sum, it will be seeking to achieve when actual negotiations begin.

Shippers will not usually find this a particularly congenial task. It has already been mentioned that the transportation shipper does not, in fact, want to purchase the services carriers have to sell; rather, what he wants is the *results* of that service, without caring a great deal by what means those results are achieved. When it comes to developing and articulating his objectives prefatory to negotiating with carriers, however, although his focus will remain results-oriented, he will find himself having to delve into the operational side of transportation.

Especially in the marketplace-focused environment of today's transportation industry, failure to scrutinize operational considerations thoroughly can be crippling to the shipper's chances of negotiating success. The range of options and alternatives available to accomplish any given set of results is simply too great, with too widely variable consequences (in terms both of price and of the probability of achieving those results consistently and reliably), to be left entirely to carriers' discretion.

This does not, of course, mean the shipper must be prepared to tell carriers how to do their job, second-guessing their operating decisions at every step of the way. But it *does* mean his negotiating position must be developed in full detail, with close attention to the process as well as the desired results of the transportation service he wants and needs—that, in sum, he needs a detailed, not merely broadly generalized, set of specifications for what he's after.

Characteristics of the Traffic

The necessary starting place is to define, as specifically as possible, the characteristics of the traffic itself—what's being moved, where it's being moved (from and to), and in what quantities. The objective here is to look at the traffic, as much as possible, through the eyes of the carriers with whom the shipper plans to negotiate, so that the negotiations may be based on common premises and understandings.

The "what" involves the *transportation* (not the manufacturing, sales, etc.) characteristics of the goods to be moved—such factors as these:

- *Commodity type and/or classification.* Standard Transportation Commodity Code (STCC) designations, ratings in the (rail) Uniform Freight Classification or the National Motor Freight Classification and other such information should be thoroughly researched to determine how the particular goods in question, and/or other, similar goods, are described and considered for transportation purposes.

• *Form of packaging.* Are the goods to move in packages or in bulk? If in packages, will they be palletized or otherwise aggregated for shipment? Does packaging meet carrier-established standards for protection of the shipments? And so forth.

• *Handling characteristics.* This applies, of course, only where the carrier will be expected to load, unload or otherwise handle the goods; otherwise it will not be part of the negotiation. If mechanical or other aids are needed for handling, or if handling requires special care, this should be noted.

• *"Stowability."* How readily will the goods fit into the fixed (usually rectangular) space and dimensions of transportation equipment? This is especially relevant where the goods are of irregular shape, or have awkward protrusions, or are too frangible for "top-loading" (that is, cannot safely have other goods stacked on top of them); such factors mean carriers will be unable to make full use of the cubic capacity of their equipment, and will thus be of considerable consequence in the negotiations.

• *Density.* Although in most cases "cube" (physical volume) will be the limiting factor on how much freight can be loaded into transportation equipment, carriers normally base their charges on weight (since weight is easier to measure). Accordingly, shipment density may be of considerable importance, especially in terms of setting minimum-weight requirements.

• *Temperature or atmospheric protection requirements.* Some goods need to be held within specific temperature ranges (frozen goods, chilled meats or dairy products or produce, certain drugs and medications, etc.). Others, while less delicate, must still be protected against extremes of heat and/or cold (especially against freezing). And certain commodities are exceptionally sensitive to excess humidity and/or moisture condensation. Special, and more costly (in terms of both capital investment and operating expenses), carrier equipment must be used to transport such goods.

• *Hazardous nature of the freight.* The term "hazardous"

here is not used in its ordinary sense, but rather refers to commodities so designated by the U.S. Department of Transportation. The DOT's list of such articles is extensive, and includes many materials not normally considered dangerous at all, such as simple aspirin tablets (poisonous in excessive dosage), common paint (flammable at high temperatures), etc. Special documentation and (sometimes) placarding of transportation vehicles hauling DOT-defined hazardous materials is required.

• *Special in-transit security requirements.* This refers to anything from simple serially numbered foil seals to cable-lock seals all the way to armored vehicles accompanied by armed guards.

• *Value.* This affects the economic risks carriers incur in taking custody of the goods, and is usually measured on a per-pound basis.

• *Fragility.* Obviously, the primary concern here is the potential for in-transit damage. In addition, however, certain especially delicate articles (electronic instruments and equipment often fall in this category, for example) are so susceptible to injury merely from the normal vibrations of the transportation process that they must move in special "air-ride" equipment.

• *Susceptibility to theft and pilferage.* Are the goods attractive to thieves in terms of their value, disposability, etc.? Can they be easily removed by the casual thief? What precautions, if any, has the shipper taken to reduce the risk of theft? (A shipping carton identified on its exterior as containing expensive watches, radios, TV sets, etc., might as well be emblazoned instead with the label "Steal Me!", as one example.)

• *Susceptibility to contamination and infestation.* Can the goods be injured by being mixed with or exposed to "foreign" materials? (This isn't always obvious; for instance, even the invisible fumes exuded by detergent powders can rapidly cause deterioration in rubber products exposed to them.) Do they attract vermin—insects, rodents, etc.?

Some of these considerations may not apply to particular traffic, whereas others not listed may come into significant play. In general, the goal is to separately identify any characteristic of the freight that will have a measurable impact on the carrier's costs of transporting it (the term is used generically here to cover not only out-of-pocket outlays but also such things as dedication of labor, vehicular capacity and other carrier resources).

Specifying the "where" of the traffic—its geographic characteristics—will normally be a fairly straightforward matter, involving merely identifying origins and destinations. Occasionally, however, a little more detail will be required because of unusual circumstances attendant to particular moves.

Some large industrial facilities, for example, have two or more separate locations for receiving or dispatching freight; especially where individual shipments may require loading at more than one such location, this must be clearly defined. This consideration should be given special attention when one or more origins/destinations is, for example, a construction job site, where delivery locations may be changed from time to time as the construction progresses. Similarly, if shipments are to move from or to sites with unusual access problems—such as center-city locations with severe traffic congestion (and perhaps local ordinances restricting traffic or parking at certain times), residential neighborhoods, rural areas accessible only by dirt roads, etc.—this should be carefully noted.

Where air or water service (or, very occasionally, rail service) is involved, it will originate and terminate at predetermined points (airports, wharves, rail team tracks) to and from which the goods must be hauled. Sometimes the shipper will have a choice of two or more such "origins" and/or "destinations"; and his selection could significantly affect the negotiations. In particular, the modern "hub-and-spoke" route structures of most airlines may make it desirable to haul air freight overland from or to a "hub" airport rather than shipping or receiving it at a nearby smaller facility served by air only via that "hub."

There remains the matter of traffic volume, which must be subdivided into two headings: Considerations that have operational implications, and those that are (from the carrier's perspective) marketing-oriented.

Under most circumstances, only per-shipment volume will

directly affect carriers' variable operating costs. Carriers can usually realize significant economies of scale only on a per-shipment basis, and thus the number of shipments and/or total amount of tonnage to be moved over any extended period of time is operationally inconsequential. The only important exception to this rule is where volume over a period of time is heavy enough to warrant dedication of equipment (particularly specialized or specially outfitted equipment), which can be used only to handle this particular traffic; and even in such cases significant scale economies will not always result.

The shipper should not overlook the cost implications of per-shipment volume in developing his negotiating posture, and should obviously seek to maximize the volume of individual shipments. At the same time, of course, he must also give consideration to other, non-transportation factors which may mitigate in the opposite direction (that is, toward smaller, not larger, shipments), such as customer order size, inventory carrying costs, etc.

Where these factors are in serious conflict, there are a variety of approaches that may be used to reduce or eliminate the so-called "small-shipment problem." The most obvious is freight consolidation—amalgamating two or more smaller shipments moving from the same origin to the same destination into a single larger one, rather than dispatching them separately. This can also be accomplished even where the origins/destinations of the individual shipments are not identical, but are geographically fairly close, by means of such devices as stopping-in-transit, assembly-and-distribution services and the like.*

Another often effective approach involves variations on what is termed "aggregate tender" of freight. In motor and air service this entails giving the carrier two or more small shipments (going to different destinations) at a single time, thus reducing the costs it incurs at origin. A similar approach is frequently used in rail

---

*For a thorough discussion of the various consolidation options and alternatives that may be employed, see *A Guide to Freight Consolidation for Shippers,* by M. J. Newbourne (Washington, DC: The Traffic Service Corp., 1976).

service where shippers have the carrier "switch in" two or more empty cars at a time, and then similarly release the cars together after they have been loaded (with goods moving, again, to different destinations).

Clearly, if the shipper is going to negotiate based on such conditions it is up to him to define them in advance, and to make certain of the practicability of meeting these shipment-volume requirements. Since the negotiated agreement will generally be predicated on the volume characteristics identified by the shipper, a realistic definition is obviously critical.

Even though carriers normally gain little in the way of cost advantages from high traffic volume over extended periods of time (as opposed to per-shipment volume), there are often marketing considerations that encourage them to give as much or more weight to this question in negotiations. It is not uncommon that a shipper with large capacity-load shipments that move irregularly or infrequently will be unable to negotiate as favorable an agreement as another who ships in smaller individual units but in overall greater quantity.

The reason for this seeming paradox is to be found in carrier accounting ledgers. Carriers of all modes have a relatively high capital investment in equipment, terminals and other tangible assets of their industry; in addition, they incur many fixed expenses—payroll, administrative facilities, permits, property taxes, etc. These costs will remain relatively constant whether or not the carrier is moving freight; they can be altered (if at all) only over relatively extended time periods and/or with considerable advance notice.

Accordingly, one of the greatest concerns of any carrier must be to keep these capital and "overhead" (the term is used loosely here) assets productive—that is, generating revenue. Carriers are therefore strongly motivated to seek the security of volume commitments over extended time periods from their customers, and are often willing to pay for those commitments with substantial discounts and rate reductions.

Periodic-volume determinations confront the shipper with a serious dilemma in most instances. On the one hand, the more volume he can promise a carrier the better will be his negotiating posture. On the other, efforts to peek into the future to ascertain how the shipper's own business will fare—and, consequently, how much or how little traffic he will be shipping and/or receiving—are obviously fraught with uncertainty.

If the traffic in question has a reasonable history, this will of course be of considerable assistance in helping the shipper anticipate how much will move in any future period of time. Corporate marketing forecasts, production/manufacturing plans and the like will be other valuable measures. Even then, however, great care must be taken in approaching this question as part of the transportation negotiation process, since none of these considerations offers anything like certainty.

As a rule, it will be best to develop both an "optimistic" and a "worst-case" scenario concerning future traffic volume. The former should, obviously, be emphasized during at least the early stages of the negotiating process, in order to excite the carrier's interest in the traffic. (At the same time, the shipper should not let optimism run amok—or, worse yet, intentionally seek to deceive carriers as to anticipated volume. Not only is such behavior unethical, but it can easily boomerang when carriers condition agreements on volume levels that are out of realistic range.)

As negotiations progress, however, it is the worst-case scenario that should fix limits on the degree of volume *commitment* the shipper will offer. Where volume-based incentive rates or deficit-weight penalties are involved (as they often will be), any agreement should obviously be predicated on a reasonable assurance, rather than a mere hope, that the shipper will have the committed volume of traffic to move.

A further consideration here is the extent to which the shipper is prepared to commit himself to any single carrier or routing for his shipments. In reaching his determinations as to traffic volume the shipper should consider how many carriers he wishes to have available to provide the service in question, and how he plans to allocate the traffic among them.

The percentage-of-traffic approach (*see Chapter 4)* has won considerable favor lately as a compromise between these two positions. Volume commitments and incentives are pegged to a percentage of a shipper's total traffic (or specified types of that traffic), rather than to fixed tonnages, thereby protecting the shipper from unexpected inability to meet forecast volumes. There are, of course, obvious difficulties inherent in such an arrangement; in particular, the shipper must be prepared to open his order books to the carrier's inspection to at least some extent. But an agreement can be worked out for this information to be held in strict confidence, and many shippers and carriers do business comfortably under arrangements of this type. Especially where future volume is highly uncertain, the shipper may want to keep this option in the back of his mind as he considers the question of volume.

The Rate and Service 'Package'

Although they're ordinarily termed "rate negotiations," that phrasing is actually a misnomer; negotiations between shippers and carriers will always involve both rates *and* service.

Many shippers make the mistake of focusing their negotiation efforts primarily or solely on price. The lowest rate, in other words, takes the freight. That's perfectly good (if trite) rhyme, but almost always bad business. No shipper seeking to compete in the marketplace economy of the United States can afford to ignore the question of freight charges; but, equally, none can afford to ignore the level of transportation service he will receive.

Yet it is not uncommon for shippers to occasionally lose sight of this. Regulation has historically worked to more or less standardize transportation service, so that the principal issue in shipper-carrier relationships of the past was strictly the rate to be paid. This standardization of service, however, no longer applies in today's deregulated industry. Carriers are now permitted, even encouraged, to offer different qualitative levels of service, to which different rate levels will apply. Shippers are thus obliged to clearly define the level of service that is being considered before they even approach the matter of rates.

The single most important service-related factor will ordinarily be transit time—not only the length of time it will require to move goods from origin to destination, but the degree of certainty with which on-time delivery (within the stated parameters) may be expected.

This question has become especially critical in recent years as the "just-in-time" approach to inventory management has been increasingly adopted by U.S. businesses. Maintaining goods in inventory costs money—in terms of both the capital tied up in on-hand stocks and the costs of building and operating (or leasing) the facilities or space for storing those stocks. "Just-in-time" contemplates the on-premises arrival of goods within a matter of minutes, hours, or (at most) a few days of their being needed for the assembly line, to meet customer orders, for display and sale, etc., thereby minimizing both the volume of inventory and the amount of inventory holding time.

But "just-in-time" also clearly requires the close cooperation of all involved in the entire procurement or distribution chain—including, of particular pertinence to the theme of this book, the carrier(s) providing the transportation service.

In this context the actual amount of transit time involved must take a back seat to reliability and consistency. If, for example, goods take an extra day in transit, the shipper/receiver for whom the carrier is providing the service may incur an additional day's cost based on the capital value of the goods (or he may not, depending on terms of sale and other factors). But that is the absolute worst he faces; and he doesn't even have to pay for storing those goods during that additional day, since they're reposing in the carrier's equipment.

But if the goods fail to show up when expected, the consequences may well be *much* more severe. Sales may be lost, assembly lines may be forced to shut down, and so on and so forth. Nor will the shipper generally have much recourse to recover the perhaps major costs of such problems from the carrier; although carriers are ordinarily liable for in-transit delay, (1) a delivery that's only a day or two late may not qualify, since carriers are normally obliged only to deliver with "reasonable dispatch" and not to meet specific time schedules, and (2) in any event the carrier will not normally be liable for "special" or "consequential" damages, such as those lost sales, the costs of shutting down the assembly line, etc.*

On the other hand, there will be circumstances under which the amount of time consumed in transit is at least as important as meeting predetermined transit schedules, and possibly more so. Fresh meats, dairy products, farm produce and other short-term perishables may deteriorate in transit and thus diminish in value; high-value goods may generate serious capital costs while in transit; in yet other cases goods may be needed simply on an "as-soon-as-possible" basis at destination; and so forth.

More rarely, transit time will be important in the opposite direction—that is, slow movement will be preferable to fast. Goods may be produced on a seasonal or other peak/valley basis, for example, to the point where they tend to overflow storage capacity at destination. In such circumstances the longer the goods remain in the transportation pipeline the less pressure is exerted on the storage facilities (since the transportation equipment serves as a *de facto* substitute for destination warehousing while the goods are in transit).

In yet other cases *neither* actual transit time *nor* scheduling reliability may be of great significance. This may be so where the goods involved are of very low value, may be stored out-of-doors or in low-cost facilities where ample space is always available, and will not be needed for commercial purposes (as raw materials, for sale, etc.) for a long time.

Whichever of these is the case, it will obviously color how the shipper approaches the negotiation process. Thus, it is critical for him to determine in advance of the negotiations how he views the matter of transit time and scheduling, and the level of importance he attaches to these aspects of the service for which he is negotiating.

---

*For a much fuller discussion of this question of carriers' legal liability for "special" damages, see *Freight Claims in Plain English,* 1982 ed., by Wm. J. Augello (Huntington, NY: Shippers National Freight Claim Council, 1982), esp. pp. 153-173.

Second only in importance to transit time—and, in fact, related to it—is the question of when, and on how much advance notice, empty carrier equipment can be made available for loading at origin.

It benefits the shipper little to have expedited transit if he must wait for extensive and/or unpredictable times before he can tender his shipments to the carrier for movement. Whether the extra time or scheduling problems occur while the goods sit at their origin or while they're aboard the carrier's equipment, the net result is the same.

The first question the shipper must ask himself in this regard is how far in advance he can be confident of planning his shipping schedule. In some cases plans can be laid many days, even weeks, ahead of actual shipment; in others shipments will be moved on an as-needed or as-ordered basis, allowing for no more than a few hours'—at most a day's—advance notice to the carrier.

Once again, the shipper must reach accurate determinations on this question before entering into serious negotiations. For most modes it is possible to develop answers for all contingencies. Rail switching schedules, motor carrier pickup arrangements can be made, air cargo space "blocked," etc., well in advance, or even set up on a regular schedule, if the shipper can count on advance planning. Alternatively, the shipper may want to negotiate for "spotted" motor carrier equipment (so that at least one empty trailer is always on-premises) or reach similar short-notice agreements with carriers, if his scheduling is unpredictable. But, obviously, he must know which is the case before he commences negotiations.

Especially careful attention must be given this question of carrier equipment availability where traffic has a highly seasonal character. Railroad covered-hopper equipment is always in short supply during grain harvest seasons, household goods carriers are chronically overbooked during the summer months, and so forth. With traffic of this sort, the shipper may want to seek especially strong commitments from carriers to ensure that adequate equipment can be obtained on a timely basis.

More prosaically, who is to load and/or unload transportation equipment, the carrier or the shipper? Generally speaking, this will have significance only where motor or (in some instances) water carriage is involved; as to railroad movements it is usual for shippers to handle all loading and consignees all unloading (although there are exceptions to this rule), whereas airlines invariably load shipments aboard and unload them from their aircraft. The shipper should reach this determination ahead of time—or, if the question hinges on cost/benefit questions associated with individual movements, determine the maximum amount he is willing to pay for carrier service rather than doing the work "in-house."

A cautionary note must be added here concerning shipments that require mechanized loading/unloading, such as by means of forklift equipment, etc. If the carrier furnishes such equipment, or expressly agrees to have its employees operate equipment provided by the shipper, all is well; but if the shipper induces carrier employees, without the clear agreement of their employer, to run the equipment, the law provides that they do so as the *shipper's,* not the carrier's, "agents"—that is, the shipper and not the carrier is legally responsible for any accidents, injuries, etc., that may occur. Since the shipper will not have had the opportunity to train carrier employees in the use of such equipment, nor even to inquire seriously concerning their skills in its use, he is taking a considerable risk in such circumstances.

As to motor and rail service, regardless of who is to be responsible for loading/unloading the shipper should attempt to develop clear information as to the amount of time these activities will normally consume (taking into consideration expectable delays, interruptions, etc.). Both modes allow only limited amounts of "free time" for loading and unloading, and assess demurrage (for rail equipment) or detention (for highway vehicles) charges for any excess. This is most critical in the case of motor carriage where the driver and power unit are expected to remain with the trailer during loading/unloading, since free time will be but a few hours at most, and detention charges will accumulate thereafter at 15-minute intervals. Rail receivers should also give particular consideration to whether it may be necessary to have incoming loaded cars "constructively placed" (on trackage not directly accessible for unloading), since free time commences to run from the date of such constructive placement even though unloading can't physically be accomplished then.

Beyond these basics, the shipper must now consider the extent (if any) to which he may need any of an extensive array of ancillary or "accessorial" services, privileges, options, etc. The following does not purport to be anything like a complete list, but is offered merely as guidance for the shipper's thinking:

- *Diversion and/or reconsignment privileges.* "Diversion" involves a decision to change the originally selected destination of the shipment before it arrives there; "reconsignment" involves re-forwarding it to a new destination after it gets to the original one but before it is unloaded. As a general rule such privileges will be used only in unusual circumstances; but in some instances they may be required on a regular basis. For example, in a few industries shippers will routinely dispatch unsold goods to arbitrarily selected points in the geographic area of expected sale, make the actual sale while the goods are *en route,* and then divert them to their now-known destination. Likewise, some distributors will order goods from suppliers in anticipation of a sale, and then, by the time delivery is made, have sold them to a customer of their own to whom the shipment is reconsigned.
- *Stopping in transit.* The most common reason for this service is to complete loading of, or partially unload, consolidated truck shipments having multiple origins and/or destinations. A variation on this type of service is the "peddle-run" delivery operation, where as many as 20 or 30 (or more) stops at delivery sites may be required. In other instances stopping in transit may be required to allow for requisite or desired in-transit inspection services of various sorts, or for other, similar reasons. Shippers of kosher meats, for example, must often have their shipments stopped in transit to meet religious requirements for preserving the kosher status of the meat.
- *Transit privileges.* This differs from stopping in transit in that the shipment is completely unloaded, is subjected to some form of processing, repackaging, etc., at that point, and is then

re-forwarded in its transmuted form to a further destination via the same carrier. Its most common historical usage has been in connection with rail shipments of grain which are stopped at a "transit point" for the grain to be milled into flour, and then moved onward (as, of course, flour). In most instances there is not even a requirement that the same physical goods comprise the inbound and outbound shipments; the goods are regarded as "fungible," and transit privileges are accorded based solely on the weight of the inbound loads. Under deregulation there has been a considerable reduction in the use of transit privileges by railroads, and it was never very widespread among carriers of other modes; but in some instances it may be of significance to a shipper.

• *"Inside" delivery service.* Where motor carrier shipments are consigned to retail establishments, offices, private residences, etc.—or even, in some cases, industrial facilities—the actual delivery location may be some distance away from the nearest point at which the truck can park (usually, although not always, inside a building). Special arrangements must generally be agreed on before the carrier's personnel will provide more than "tailgate" delivery.

• *Sorting and segregation of freight.* Again, this normally applies only in the case of motor carrier shipments, and involves additional work on the part of the driver at destination to sort or segregate shipments of mixed articles.

• *Loading/unloading equipment for bulk commodities.* Various types of valves, connectors, tubing, hoses, etc., will be required to interconnect between bulk equipment of transportation carriers and the shipper's and/or receiver's storage tanks. The carrier will not normally have such equipment available unless it has expressly agreed to do so.

• *Return of used pallets, shipping containers, etc.* Where the shipper employs reusable packing or shipping materials in preparing his shipments for transportation, he will, obviously, want them back. If the carrier is expected to return them, or to be financially accountable for their loss if they're not returned, this will have to be given attention.

• *"Turnaround" service.* In the usual course of events carrier equipment will arrive loaded at destination, be unloaded, and depart empty. In some instances, however, shipper and consignee will have a steady flow of two-way traffic between them which can readily move in the same equipment—or the consignee may have loads destined to other shippers for which that equipment can be used. Since delivery of the inbound load and pickup of the outbound one is accomplished at a single stroke, the carrier saves appreciable operating costs in such circumstances, which may provide an important negotiating point.

• *"Special" damages.* As noted (*see above*), transportation carriers are not normally liable for more than the actual injury *to the goods themselves* resulting from any in-transit loss, damage or delay. Shippers may only collect for other forms of economic injury that are the consequences of the loss, damage or delay—such as enforced worker or equipment down-time at destination because needed goods weren't available on time, etc.—if they give carriers advance notification that such "special" damages may result. Normally speaking, the presence of such a warning on the shipping documents causes the carrier to immediately lose all interest in hauling a shipment; if the shipper expects the carrier to assume such additional liability, he should include this in the negotiations.

• *C.O.D. shipments.* Most business transactions in today's world are handled on the basis of trust; either the shipper forwards the goods to his customer under credit arrangements, or the customer makes payment in advance and trusts the shipper to fill the order. On occasion, however, it will happen that the parties don't choose to do things this way, but rather to exchange goods and money simultaneously at the point of delivery. As to such collect-on-delivery (C.O.D.) shipments, the carrier will usually serve as the shipper's agent in collecting the consignee's payment before unloading takes place.

Once again, it must be stressed that the foregoing does not purport to represent a complete or comprehensive listing of all of the service options and alternatives available. The shipper must determine for himself, case by case and (in as much detail as he can), precisely what service he is seeking—and what will be the consequences if he must accept a lesser level of service. This amounts to the strategic planning that is a prerequisite to the success of any negotiation process.

Such planning will normally involve consultation among several departments within the shipper's organization. The Traffic Department must, of course, be involved, since it will normally be the principal representative of the shipper in its dealings with carriers. On outbound movements, the Sales/Marketing Department will also have a major input; it is, after all, most directly concerned with customer service, which in turn depends in considerable part on the organization's transportation arrangements. If inbound traffic is at issue, the Procurement and/or Production Departments should certainly participate, since deliveries of the supplies and/or raw materials the company is receiving will be crucial for them. The Accounting Department may be concerned about freight bill payment and auditing; Legal will have something to say about transportation contracts; Inventory Control about transit times and pickup/delivery scheduling; and so forth.

At this stage the shipper need not be overly "realistic" about his goals. The negotiations process itself will introduce all the "reality" he could ask for—more, very likely, than he wants. There's no need to anticipate that much-later stage yet; the focus of the initial planning process should be mainly on development of a "wish list" representing the ideal of what the shipper would like to see in the final agreement.

(Of course, it's possible to carry unreality too far. Certainly all shippers would love to have their goods hauled for nothing; but that's not remotely likely. On the other hand, the shipper can certainly shoot for rates, say, 40-50% (or perhaps even more) below present levels; while most likely unattainable, this is at least somewhere within the realm of possibility. The idea is to come as near the edges of the possible as one can without overstepping the borders into Never-Never-Land.)

Once this "wish list" has been finished, the next step in the shipper's strategic planning process is to turn things around 180°

to develop the opposite side of the coin—what might be called a "worst-case list." The shipper knows the ideal of what he's seeking—now what's the very least he's willing to settle for?

In most cases the parameters will not vary uniformly. Indeed, as to some of his service objectives the shipper may not be willing to come down at all; the ideal of what he wants is also the bare minimum he needs. In other areas, however, there may be room for a great deal of negotiating flexibility.

The process of negotiation will necessarily be one of give and take. In such an environment it's obviously advantageous to know going in just how much "give" there is in one's initial position—what can be traded off, and to what degree, and what is non-negotiable.

A further step down this path is prioritization. The company is obviously going to try to come as close as possible to its ideal (and stay as far away as possible from its worst-case level of minimum acceptability). So what would it prefer to have its negotiators trade off first, and what should they cling to until and unless they *must* give way? It's obviously better to make such decisions in advance, when there's time to think and plan (and, to the extent necessary, consult others in the company) than to wait until the last minute, in the midst of actual negotiations, and then "wing it" with spur-of-the-moment decision-making.

The Question of Pricing

Thus far, it will be noted, there has been no mention of price. This is because what one pays for something is obviously a function of what that "something" is, so that questions about the pricing of transportation service cannot be addressed until the specifications of the desired service have been completed.

Many shippers are prone to take a somewhat vague attitude about their pricing objectives, simply negotiating in general terms for the lowest rate level they can get—about the same way the consumer goes bargain-hunting in local shops. And indeed there are circumstances in which this will be the desired approach; but such circumstances will normally best be served by the techniques of competitive bidding (*see Chapter 9),* rather than direct negotiation.

For the most part, the shipper should seek to articulate his pricing objectives as carefully, and in as much detail, as those related to service. He should once again identify an optimal desired ("wish-list") level of rates and charges, and a worst-case "ceiling," before he enters into negotiations.

The obvious starting point for these determinations is the level of rates the shipper is now paying for transportation of either the same or (if the traffic in question does not have a transportation history) similar traffic. It is extremely important that this information be gleaned from actual movements, not a mere perusal of carrier tariffs, rate circulars, etc. In the present era of discounting, contract service, etc., it is often difficult or impossible to ascertain from such documents the real-world rates and charges under which traffic is actually moving; the risk of uncovering mere "paper rates" through such tariff research is thus substantial.

Additionally, the shipper should also seek, to the maximum extent possible, to ascertain how much his competitors, and/or others with similar traffic, are paying for *their* transportation service. This will rarely be easy, and it may often prove impossible to develop more than educated guesses (if even that) because, again, published tariff information may not always reflect real-world pricing. Even information extracted from carrier freight bills (as when, for example, the shipper compares his traffic to that of others who are shipping to him under agreements they have negotiated with carriers) may not paint an accurate picture; in some instances carriers grant shippers discounts, allowances, etc., that are paid separately and not shown on actual freight bills.*

A third obvious area for research—yet one that is often

---

*It should be noted that this may be the case irrespective of who pays the freight bill; that is, discounts, allowances and the like are legally payable to parties other than the actual freight bill payor. This question is discussed in more detail below.

neglected by shippers—is carrier costing. The reason for this neglect lies in the fact that in other economic sectors (the sectors whose procurement practices are increasingly being used as a model for the purchasing of transportation services in many organizations) such cost information will be the carefully guarded secret of the vendor, so that purchasers can rarely develop cost information within even "ballpark" ranges.

Here the regulated status of most transportation service (even in its presently limited form) stands the shipper in good stead. Although such requirements have been seriously weakened in recent years, regulated carriers are still (as of the time this was written) obliged to file periodic reports with regulatory bodies which include considerable detailed cost information; and these reports are, by law, publicly accessible. Even where individual carriers may not be subject to such reporting requirements, the agencies maintain "regional average" cost information that can serve as an acceptable (although admittedly less precise) substitute. And finally, the regulatory agencies have obligingly developed costing models—such as Highway Form B for motor carriers, Rail Form A or the newer Uniform Rail Costing System (URCS) for railroads—which allow specific application of this cost information to particular movements.

It must be emphasized that such cost data will not always, or even usually, coincide with a given carrier's determination of its own costs. In the first place, the carrier will have available much greater cost information than it is obliged to report to regulatory authorities. Second, transportation cost-finding is far from an exact science; there are many different bases on which costs may be allocated to particular services or activities, and not all carriers approach such questions identically. In particular, the standardized costing models developed by the regulators are rarely used by carriers for their own internal accounting purposes.

Nevertheless, the shipper who fails to look as closely as circumstances will allow at the question of carrier costs is depriving himself of a potentially important negotiating "edge." It is obviously a big help to anyone entering a business negotiation to know in advance roughly where the other side's break-even point lies; and, unlike in other economic sectors, this information can often be developed within fairly close tolerances in transportation.

(A cautionary note: Shippers unfamiliar with transportation costing are probably best advised not to attempt this intricate process themselves, but to engage professional consulting firms who specialize in this work. The process of costing is both difficult and conducive to errors—even a small mistake at a key stage can skew the results radically—and experience and expertise are much to be prized in this endeavor.)

In addition, the shipper should also explore the "value of service" issue from his own perspective—what's the service worth to him? An obvious question to be researched in this context is how much it would cost him to provide the service on his own, rather than purchase it from outside vendors.

Where motor carrier service is involved—or service that *can* be feasibly handled by motor carriage, even if negotiations with carriers of other modes are contemplated—this question can encompass the entirety of the service. It should be evident that the cost of private truck service should form the absolute price ceiling during any negotiation with for-hire carriers.*

Even where full private carriage is unfeasible (operationally or economically), it may be that certain portions of the transportation service desired can be provided by the shipper as well as the carrier; a common example is loading/unloading service. In such cases these particular services should be segregated for purposes of the overall negotiation, so that a rational cost/benefit decision may be made on them as a separate increment.

---

*Shippers' decisions to start or withdraw from private trucking operations also involve a wide range of other factors, many of them non-economic in nature; but since the comparison here described is being drawn purely to set limits for negotiations with for-hire carriers, such factors may generally be disregarded for this limited purposes. For a more comprehensive discussion of the role of private carriage, see the author's *Practical Handbook of Private Trucking* (Washington, DC: The Traffic Service Corp., 1983).

The cost/benefit approach should also be rigorously applied to any area of service where there is significant negotiating flexibility between "wish-list" and worst-case levels. The trade-offs that will take place during negotiations will generally involve both service and price, and it is obviously desirable for negotiators to know approximately what trade-offs are favorable and what are not—to know, say, that one-day-shorter transit time is worth 5¢ a hundredweight in additional freight charges, but not 7¢.

Finally, in some instances it will be necessary to take still a different approach to the matter of "value of service"—at what pricing juncture does the negotiation become moot? Transportation accounts for but a small proportion of the market cost of most goods—too small for pricing variances to have a significant impact on overall market position. But as to certain products which are relatively low-priced or which must compete in extremely price-sensitive markets, this will not be the case; transportation costs may be a "make-or-break" factor under these circumstances. If this is the case, negotiators must be acutely aware of precisely how, and to what degree, transportation costs will affect their markets in order to negotiate intelligently.

The Carrot, the Stick and Other Matters

A final feature of the negotiations process on which the shipper should develop a clear posture involves penalties, incentives, discounts, allowances and other such contingency features of the ratemaking process.

To begin with, it should be obvious that every commitment the shipper seeks from the carrier must, to be meaningful, be backed up by economic considerations. The carrier must have some form of economic stake in fulfilling its promises; otherwise the promises stand as mere words, unenforceable in any serious way by the shipper and easily forgotten or neglected by the carrier. In other words, to be worth significant attention during negotiations, a commitment should have "teeth" in it.

There are two ways to approach this question—by means of

incentives that economically reward the carrier for on-target performance, or through penalties that deprive it of revenues if it falls short of the mark. Any service element that the shipper deems of key importance should generally be supported by one or the other (or some combination of both).

Among other things, broaching the idea of incentives and penalties during the course of a negotiation will quickly bring discussions down to earth. Lavish promises concerning the calibre of service the shipper may expect will quickly give way to hard-boiled assessments of what he can *really* count on; service elements that were airily dismissed on a "that-goes-without-saying" basis will be instantaneously resurrected as matters worthy of serious consideration. As any child learns almost from the cradle, a promise is one thing when it's merely an expectation and quite another when it's backed by threats of punishment or withheld treats—a lesson that applies equally to the schoolyard and the business office.

At the same time, a realistic, punishment-fits-the-crime approach should be taken in this area; neither incentives nor penalties (whichever is used) should be inappropriately out of proportion to the relative importance of the service element in question. If the level is set too low the carrier's incentive to meet its commitments is blunted; if it's too high, the shipper may find that he actually reaps a "net profit" from carrier service failures and so is motivated to surreptitiously encourage them. Once again, rigorous cost/benefit analysis should be applied to such questions.

Are incentives or penalties more effective? The question is imponderable, dependent more on the mentalities of the parties involved than on any objective factors. Some people respond best to the carrot, others to the stick, and some need both; decisions as to which is superior in a particular situation should be decided within the context of that situation, rather than as a matter of preordained policy.

It should be noted, however, that each approach communicates a subliminal message to the carrier, which may in some instances be of importance. Use of the incentive says the shipper recognizes that a lower level of service than what he wants is actually standard, and that he is asking for, and proposing to reward, above-standard service. The penalty system, by contrast, conveys that the wanted service is indeed the standard, and that it is therefore appropriate to penalize failures. For want of a better guide, the choice between incentives and penalties can be based on how the shipper's particular service needs compare with the "norm" of service the carrier provides its customers generally.

If the shipper is going to raise the incentive/penalty issue, of course (or even, in some cases, if he doesn't), he must be prepared for a response in kind from the other side of the bargaining table. That is, to the extent the shipper is making commitments of his own—most often, though not always, in connection with traffic volume—he must expect the carrier to press for incentives or penalties to enforce those commitments on him. Clearly, few carriers will be prepared to regard this issue as a one-way street, and the shipper should enter the negotiation with a clear idea of how far he is prepared to go, and with what economic considerations at stake, in this regard.

The modern-day practice of using discounts, allowances and the like to reduce freight bills also warrants some attention by the shipper. In the majority of cases this form of ratemaking is employed simply as a convenience, allowing carriers to reduce their rates below existing published levels (such as bureau tariffs) without having to go through the wearisome process of republishing, rate for rate, the often highly complex existing scales. Alternatively, discounts/allowances may be used to reward shippers for particular activities—performing loading/unloading, meeting defined volume levels (a variation of the incentive/penalty theme), etc.

As a rule, such discounts and allowances are deducted on the freight bill itself, so that only a "net" amount (with the discount/allowance factored out) is billed and paid. This obviously helps both parties minimize bookkeeping requirements, rather than obliging them to unnecessarily exchange checks (the shipper paying the undiscounted charges, the carrier refunding the discount, etc.).

For various reasons, however, some carriers and shippers do things otherwise. In one variation, a freight bill is submitted for the undiscounted charges but the shipper's accounting department lops off the discount before paying. This preserves most of the bookkeeping advantages discussed above, but almost always heralds some form of clandestine dealings, the ethics (and even the legality) of which may be highly dubious. In another, the exchange of checks described above actually does take place, which may or may not involve ethical and/or legal questions.

If the exchange-of-checks method is used, it may simply represent a variant form of the incentive/penalty approach described above. For example, it may be agreed that a shipper will pay single-shipment rates until and unless his traffic reaches a certain level of aggregate volume, at which juncture he will be entitled to a refund on past shipments as well as a discount on future ones. Or any agreed penalties against the carrier for service failures may likewise be payable separately. Or, again, this approach may be used to implement a confidentiality agreement between shipper and carrier for entirely legitimate purposes.

Sometimes when confidentiality is the reason, however, the motivations are not entirely laudable. On occasion shippers will deliberately seek to conceal the existence of their negotiated discounts from their customers, to whom they are rebilling full (undiscounted) freight charges in their invoices; similarly, where terms of sale are "freight-collect-and-allowed [off the shipper/vendor's invoice]," the consignee may have similar motives. And since discounts and allowances are legally payable to parties other than the payor of the freight bill, similar situations may exist where the shipper negotiates rates on traffic that moves on a freight-collect basis, or the consignee on prepaid traffic.

Since this form of ratemaking is quite new to the transportation industry (almost all of it postdates the deregulatory legislation of 1980), there are, as of the time this was written, no clear-cut guidelines on these practices that have been established through formal litigation. Common sense, general standards of business ethics and an examination of comparable case law developed in other business sectors, however, leads to the following rule-of-thumb guidelines:

First, any discount or allowance (by whichever name) directly related to the performance of a particular activity by a shipper or receiver—such as loading/unloading, furnishing private transportation equipment, etc.—is rightfully payable only to the party that performs the activity. This is based on the same rationale that induces carriers to assess detention or demurrage charges at origin against the shipper and those at destination against the receiver, and is fairly clear-cut both ethically and legally.

Second, allowances and discounts intended as incentives for shipper or receiver likewise belong, at least arguably (this area isn't quite so clear-cut), to that party. In this category might fall such things as aggregate-tender allowances for the shipper, turnaround or reshipment discounts for the receiver, etc. Once again, it is in that party's sole discretion whether to take the action the incentive is intended to encourage, and the reward for doing so is therefore, reasonably, its to keep.

Considerably less clear is the status of the marketing-based discounts sometimes accorded to large receivers who do business with many small vendors. In order to minimize congestion at their receiving docks and otherwise simplify matters for themselves (at least, this *used* to be the reason), such receivers prepare detailed "routing guides" for their vendors identifying the carrier or carriers the should be used. Vendors disregard such guides at their peril; some shippers assess substantial dollar penalties on vendors who don't follow the required routings, while others simply refuse to accept such "misrouted" shipments.

Recognizing that the buying power actually reposes with the receiver in these cases—even though it will often be the shippers who pay freight charges—carriers will grant such receivers sometimes-substantial discounts to "pay" for having their names listed in the routing guides. Arguments can be advanced in such cases that the receivers are entitled to keep the discount payments; but especially where the shippers wind up having to bear higher transportation charges than they would via alternative routings, there are powerful counter-arguments, as well.

Those counter-arguments gain additional force where the party receiving the discount conceals it from the party paying (or bearing) the freight charges. In some industries where this problem has already aroused considerable controversy, attorneys representing the aggrieved freight-bill payors have made public statements that such "hidden" discounts run flatly counter to the law. Such statements must be taken with something of a grain of salt, considering the bias of their sources; but neither may they safely be ignored. It is evident that, at best, such discounts have an uncertain legal, and an even rockier ethical, status.

It should be emphasized, however, that these questions do *not* relate to the relationship between carrier and shipper (or carrier and receiver, or even carrier and third party, as the case may be). It is fairly well established that the carrier may pay discounts to any party it chooses, and that party may receive those discounts, within the full parameters of both legal and ethical behavior.

The questions that arise have to do, rather, with the vendor-buyer relationship. That is, having received those legal/ethical discount payments from the carrier, is the shipper or receiver entitled to *keep* the money, withholding it from some other party who has actually paid or borne undiscounted freight charges? The answer may in many instances be contingent on the specific language of sales contracts, purchase orders or other vendor-buyer agreements, and cannot be given categorically; but the question is one that should be addressed by any shipper in such a position *before* he approaches negotiations with carriers on the subject.

Common Pitfalls to Avoid

There are several pitfalls that await those who neglect this kind of strategic preparation for negotiations. The most common problem is a failure of the company to either (1) define, in clear terms, its objectives, or (2) communicate those objectives clearly and completely to the individual(s) responsible for the negotiations. Either is crippling, because it means that the company is, in essence, approaching the negotiations without a good perspective of what it wants to achieve through them.

A second problem involves inadequate attention to the "wish list," so that it becomes either too wishful or (more often) not wishful enough. The purpose of preparing such a wish list is to establish a starting position for the negotiations that are going to take place; this is the initial proposal the shipper plans to lay on the table.

If the list is too demanding, one of two things—both bad—is bound to happen: Either the carrier will simply walk away from the table, feeling there isn't a sufficient basis for negotiations that stand a reasonable chance of bearing fruit; or the over-demanding shipper will have to start giving way immediately and heavily, conveying the impression that he's wishy-washy and can be persuaded to give up a lot more as he negotiates.

Since most people have an innate psychological desire to be "accepted" by others, this problem doesn't crop up in business negotiations so often as one might expect. What happens, rather, is the opposite—an initial proposal that sabotages the shipper's bargaining position from the start by being excessively lenient.

Many people fail (or are unable) to make a clear delineation, in their own thought processes, between their "business" and "social" selves. In a social setting, of course, one's objective is to win the esteem and affection of others, and this is not furthered by pressuring or disagreeing strongly with those "others." It should go without saying that such socially oriented motivations have no proper role at the business bargaining table—especially where one does not appear there on one's own behalf, but rather only as a representative of one's employer—but this does not prevent such an attitude from adversely affecting the negotiation process in many instances.

Managerial measures of employee performance may also favor an inappropriately "soft line" approach to negotiations. In most areas of business management performance goals will be fixed with the intention that their attainment is expected, and that failure to do so is legitimately an indication of inadequate achievement. If the same standard is applied to the individual responsible for business negotiations, however, the expectation is that he will emerge from the bargaining with most or all of his original proposal intact. Since he knows he will be "marked down" by his superiors if he doesn't manage this, his self-interest obviously mitigates in favor of modifying his (and his company's) plan to something he is fairly confident he can "sell" the carrier pretty much unchanged.

Thus, fearing to ask for too much, a surprising number of shippers go to the opposite end of the spectrum and seek too *little* instead—less, in fact, that the carrier may have been prepared to concede. Since obviously one is not likely to get a better deal than one initially proposes (and probably won't do even that well, no matter how reasonable the initial proposal may be), this, too, is a mistake.

As between these extremes, if a choice must be made it's definitely better to ask for more rather than less. This is a philosophy much overworked in certain consumer sectors of our economy (the automobile industry is probably the classic example), where "list" prices are held purposefully high in order to leave room for ample point-of-sale "concessions" by retailers. But it is overworked because it's generally effective (even where the tactic is obvious), a lesson shippers might well apply to their own dealings with carriers. A basic rule of thumb should be that if the initial proposal *is* in fact accepted more or less as presented, the negotiating strategy needs an overhaul; while not always the case, this usually means the shipper could, by taking a harder line, have got an improved result.

Still another potential strategic pitfall is failing to adequately establish bottom-line parameters—that "worst-case list"—and/or (again) to communicate those parameters adequately to the actual negotiator(s).* If the negotiators don't know

*A not-at-all-apocryphal tale—this one actually happened—illustrates the importance of *clear* communications between those responsible for the shipper's strategic planning and those involved in the actual negotiations. The story concerns the training of soldiers to man the guns of bomber aircraft during

before they start talking just where the bottom line is, it should be obvious that their chances of inadvertently crossing it will be fairly great. No "deal" should be pursued beyond its predetermined merits; but if those merits aren't properly defined and quantified, following this rule will clearly be more a matter of luck than design.

Researching the Other Side

The final step—and final potential pitfall—in strategic planning is development of as good an informational base as possible about the carrier(s) with whom negotiations are planned. It's the old principle of not seeking to sell refrigerators to Eskimos; when it comes down to actual negotiations, the idea is to trade off what the other fellow wants, not uselessly give him something that's of little or no value to him—but the surrender of which may be costly to the giver.

The shipper's major concerns here should be with:

• *Carrier service reputation.* Information about this is usually obtainable from other shipper managers through informal channels. It's obviously best not to seek information from direct

---

World War II. The exigencies of wartime meant that such training had to be on simulators; no bombers could be spared from combat. But the training facility *did* have a few obsolete two-seater, open-cockpit planes available; and the training commander decided that all trainees should be given at least one ride aloft in these craft, to give them a "taste of the air."

One of the first trainees to take such a flight was obviously terrified—a terror that was only intensified when he was strapped into a parachute and given detailed instructions in its use. He clung grimly to his seat as the plane took off. Recognizing his fear, the pilot thought to distract him by giving him an aerial view of his barracks. Buzzing low, he heeled the tiny plane over on one wing for better visibility, turned around in his seat and (unable to be heard over the rush of the wind) pointed first at the trainee and then at the barracks on the ground, to show him where he slept at night. The trainee gave him one horrified look—and bailed out!

Similar communications failures in connection with the business negotiations process can produce similarly inappropriate results; and just as was the case with the trainee in the foregoing story, the humor tends to be lost to those involved.

competitors in the same industry; the reliability of such information will be open to some question. But many other options exist; and the carrier can always be asked (like vendors in other economic sectors) for references.

• *Carrier equipment.* This is particularly important where the shipper needs specialized equipment to handle his traffic—refrigerated vehicles, lowboys, covered hopper cars, etc. Does the carrier have, or have access to, an adequate supply of such equipment? Carriers are often over-optimistic about their ability to procure equipment they don't now have; but if they can't it will ultimately be the shipper, depending on availability of that equipment to move his goods, who will pay the price.

• *Operational experience and expertise.* Where the carrier is already established in the market sector in question, this may generally be taken for granted (subject, of course, to a check of its service reputation). Where this service which is the subject of the negotiations will represent a significant market expansion or change for the carrier, however—a previously package-freight carrier hauling bulk commodities, a carrier serving a new geographic area, etc.—closer attention, obviously, should be paid to the carrier's qualifications to handle the traffic effectively and efficiently.

• *Financial stability.* If the carrier "goes down" financially—lapses into insolvency and/or bankruptcy—the shipper will obviously also be the loser; his service will be instantly cut off, he can forget about pending claims or unpaid discounts, etc., etc. Even if its financial problems are less drastic, they may still precipitate cost-cutting that can impair service quality, or make the carrier easy prey for a merger or acquisition that will lead to a repudiation (by the surviving or acquiring firm) of the negotiated agreement. Especially in a marketplace as volatile as that of the deregulated transportation industry, shippers should take steps to assure themselves of the financial soundness of any carrier with whom they propose to do significant business.

• *Liability insurance.* Does the carrier have adequate coverage in the event of accidents or other untoward happenings? In the present era of "deep-pocket" jurisprudence—with judges and juries prone to award damages against defendants based as much on their ability to pay as on their contribution to the plaintiff's injury—this can be of some consequence. It is quite possible that a shipper who employed a carrier that lacked adequate insurance could be deemed negligent for having done so, and could therefore be found collaterally liable for damages resulting from an accident in which the carrier was at fault. Especially where carrier liability insurance coverage is mandated by law (as is, especially, the case with respect to motor carriage), the shipper, for its own protection, should take steps to assure itself the carrier does indeed possess such coverage.

Quite obviously, where interline service is involved, this research should extend to all of the participating carriers. Like the proverbial chain, interline service can be no stronger than its weakest link.

Finally, there is one more question into which the shipper should inquire—the "value of traffic" to the particular carrier(s) with which he plans to negotiate.

Like "value of service" from the other side of the table, value of traffic refers to the perception of carriers that some types of traffic are inherently more desirable than others. In some instances this perception will be the result of managerial prejudices or other non-substantive "intangibles"; but mostly it will derive from marketing or operational factors that the observant shipper may be able to identify.

For example, the carrier may be seeking to establish itself in a new market area, and for that reason be prepared to grant concessions beyond the norm. Or it may simply want an "in" with a large shipper, which that shipper may be able to exploit into an initial bargaining-table advantage. Or—and this is by far the most common situation from which the shipper can hope to benefit—it may have a geographic traffic imbalance which the traffic in question would help rectify.

Certain areas of the country are chronically imbalanced in this fashion, and most shippers "exporting" from (or "importing" to) those areas know full well the advantage this gives them. Little need be said about such cases; rates for backhaul service from such regions are commonly at below-cost levels as the carriers seek out any revenue they can get in preference to their option of "deadheading." But in other cases an alert shipper may be able to identify imbalanced traffic lanes that are specific to a particular carrier, and be able to use this knowledge to his advantage.

At the same time, it's obvious the value of traffic is a two-edged sword; it can work either to the benefit or the disadvantage of the shipper. Where he finds himself on the wrong side of the blade, the shipper should give serious consideration to negotiating, instead, with some other carrier or carriers where the odds are not so heavily stacked against him.

The Special Case of Captive Traffic

A good deal of the foregoing contemplates the existence of at least some degree of competition for the shipper's traffic. Unfortunately (from the shipper's perspective), such will not always be the case.

All so-called "captive traffic" in today's marketplace is rail traffic. Railroads alone, among the transportation modes, operate by means of proprietary travelways, so that competitive access to shipping and receiving points by other railroads is largely in the hands of the track-owning carriers. And since some traffic can feasibly move only by railroad, this means in some cases rail carriers will have a *de facto* monopoly.*

It may seem that all the cards are stacked against the shipper who is in this kind of negotiating situation; and there's a lot of truth in that. But the shipper may have more opportunities than

---

*As discussed in Chapter 2, there are nominal regulatory limitations on the carriers' exploitation of such monopoly situations, to which the beleaguered shipper may perhaps have recourse. These limitations, however, have been so liberally construed by the post-1980 ICC that they have been of little practical assistance to most captive shippers. A plea for regulatory intervention should always be regarded as a last, desperate resort; captive or not, it is primarily through the negotiating process that the shipper must seek his results.

he at first realizes to achieve favorable results at the bargaining table.

First, it is important to distinguish between traffic that is permanently, inescapably captive and that which is merely so for the moment. Even if there are no transportation alternatives presently available, this does not necessarily mean that the shipper cannot develop such alternatives. Coal, for example, is widely considered captive to railroads, but there are no operational reasons why it cannot be moved at least short distances *i.e.,* to points of interchange with competing railroads, barge lines, etc.) by truck. The same holds true with respect to mineral ores and, indeed, to most other commodities now popularly deemed to be "owned" by the railroads.

In this context it is worth noting that, not so very long ago, grain was likewise regarded as the railroads' exclusive province. Since then, however, motor carriers have made considerable inroads into the grain transportation market. And it should be remembered that post-1980 relaxations in highway size-and-weight laws appreciably enhance the ability of truckers to compete for such mass movements of bulk traffic.

Additionally, such alternatives as short-distance slurry pipelines, water service and the like may be potentially available to the seemingly captive rail shipper. And in some instances shippers have even acquired their own short-line railroads to avoid captive-traffic status *(see, again, Chapter 2).*

To be sure, it probably will require considerable effort, perhaps substantial capital investment (in the building of rail lines and access roads, laying of pipelines, dredging of watercourses, etc.), and so forth to make these alternatives feasible. And even then, transportation costs may rise beyond what the shipper now deems "reasonable." But it may also be that this effort and expense will never in fact be necessary; the mere potential for development of such competition, and the shipper's evident willingness to pursue such alternatives, may alone be enough to influence the policies of the rail carrier presently serving him.

Railroads prize their captive traffic greatly. Over the past several decades they have suffered dramatic erosion of their traffic, primarily at the hands of motor carriers; between 1939 and 1984, the railroads' share of "intercity" (*i.e.,* non-local) freight traffic plummeted from over 62% to about 36½%, most of which (aside from oil and oil products, which went to the pipelines) was stripped away by truckers.* More and more, captive traffic plays a major role in their future planning and security; and they are apt to take any threat to it, even if slightly far-fetched or lying some years ahead, quite seriously.

Even where such competitive threats cannot be credibly mounted—for one reason or another, the traffic truly *is* captive to the rail carrier in question—there are a variety of other negotiating approaches than can be employed.

One of these depends on whether substantially all, or merely a part, of the shipper's traffic is captive—and also on whether, as to the non-captive portion (if any), the shipper has the option of sending it, too, via the same railroad. If both questions can be answered favorably, the competitive traffic can be obviously be held out as an incentive to the carrier to "be reasonable" about the captive service. (This must be approached in fairly gingerly fashion, however, to avoid possible antitrust problems concerning "tying arrangements"—*see Chapter 10.)*

If such is not the case, the shipper has two alternatives, which will always be available. The first is to work toward a long-term contract, one of perhaps 10 or 20 years' duration and with substantial volume guarantees. Like most people, railroad managers regard the future with some trepidation; yes, the traffic *appears* to be secure right now, but who can be certain what the future might bring? The added security of an extended contractual commitment affords the shipper at least some degree of bargaining leverage—not to mention the added benefit of at least assuring that his transportation costs will not increase out of control for the duration of the contract.

---

*Figures based on ton-miles (a ton-mile is one ton hauled one mile), extracted from *Transportation in America,* 3rd ed., *op cit.*

Secondly, the shipper can open up his market forecasts to the carrier's perusal. Most rail captive traffic is (a) relatively low-value, and (b) to at least some degree sold in competition with either the same or substitutable commodities (or both). If transportation costs rise too high in such circumstances, the marketability of the materials will suffer; and as their market declines, so, inexorably, will the volume of the traffic. It is obviously not in the railroad's best interests to price the traffic so high that this sequence of events is set in motion.

If the shipper adopts this second course of action, however, he is best advised to be relatively frank with the carrier, shading his forecasts, if at all, lightly. There are many sources of information about virtually any market, and many of them are readily available to railroads; and the shipper can be sure his data will be scrutinized closely by the carrier with which he's negotiating. Significant exaggeration of his own market position will not only be quite evident to the railroad, it will destroy the shipper's bargaining credibility for the future. Indeed, if the shipper is aware that his own forecasts differ from those of independent market researchers, he should take exceptional pains to explain the rationale behind his determinations, even if this means giving away what would otherwise be regarded as highly confidential information.

Finally, if all else fails the shipper may wave the litigatory flag as a last-ditch measure—threatening to invoke regulatory authority. It has already noted that this threat will in all likelihood prove an empty one, since affirmative regulatory action in such cases has been (as of the time this was written) so rare as to be virtually an endangered species. Nevertheless (again, as of this writing), railroads were regarding the possibility of such litigation with a good deal more seriousness than might be expected by objective measures—especially where shippers were also in a position to enlist political aid for their causes. Accordingly, notwithstanding the improbability of regulatory success, it could be that such a threat would serve to stir the carrier of a previously intransigent bargaining position to at least some degree.

6

COMPUTERS AND STRATEGIC PLANNING

The advent and increasingly common use of the computer as a management tool has given both shippers and carriers extraordinary opportunities to fine-tune their strategic planning processes preparatory to transportation rate negotiations. Especially with the inexpensive desk-top microcomputer at his instantaneous disposal, the transportation manager can do his "homework" more rapidly, and with a greater degree of precision, than was ever before possible.

Traffic-lane analysis, for example, is one of the key ingredients of planning. The shipper needs to know when and where its traffic is moving, in what volumes, etc., to develop the best leverage in negotiations. The carrier, for its part, needs to understand how this traffic will complement its other operations, where consolidation opportunities may be found, etc.

There are a number of computer-assisted procedures for both to develop this information. At the "high end" of the scale it is possible to purchase computer software identifying distances (air, rail, highway or water) between any pair of points in the country, locating intermediate points, identifying off-route points where traffic can economically be consolidated, etc., etc. Software even exists to determine the most efficient highway routing between pairs of points (taking into consideration not only mileage but also speed limits, highway conditions, etc.).

This much detail, however, will not always be necessary. Often it may suffice merely to identify points of origins and destination and anticipated traffic volumes without more. Specialized software likewise exists to do this job; or, alternatively, general-purpose software such as the popular DBase II (or III) and Lotus 1-2-3 programs (and many others) can be adapted to such applications.

In this context it is also possible to make effective use of another of the computer's many potentials—its ability to translate raw data, such as traffic-lane information, into graphic depictions. Such a picture may indeed be "worth 1,000 words" during the negotiating process—or even before negotiations begin, as a means of developing support within one's own company for a particular strategic objective.

Computers can also provide a tremendous assist in *comparative rate research,* as both shipper and carrier strive to ascertain what alternatives the marketplace offers for the transportation involved in the negotiations. Rate research has always been at best a difficult undertaking; and the proliferation of tariffs spawned by deregulation has aggravated the difficulty manyfold. Today most carriers publish their own rates, usually in their own tariffs, so that determining competitive rate levels can involve slightly more effort than counting the individual grains in your dining-table salt shaker.

Enter the computer. There are a number of commercial services that maintain extensive "data bases" of rate information, via all modes of transportation. These can be accessed over the telephone using a computer and a "modem" (MOdulator-DEModulator, which translates the digital electronic pulses used by computers into signals suitable for transmission over ordinary telephone lines). Such services, of course, are not offered free; but their cost is generally quite modest by comparison with the time and effort such rate-research tasks require if performed manually.

Computers can also assist in a variety of rate comparisons involving detailed calculations, such as comparisons of the rates over multiple routes, comparison of total charges where a variety of accessorial services are involved, etc. Specialized software for these and other comparison-shopping-type applications is available from a number of sources (including some of the service agencies that maintain rate data bases, as well as others); or, again, more generalized programming can be adapted for the purpose.

One of the most important uses of computers in transportation contracting is in *cost-finding.* For shippers, software exists that can transform the data included in carriers' annual regulatory reports into movement-specific costs, or, where no annual report is available, can use regional-average information developed by regulatory bodies (as well as, in some cases, proprietary averaging techniques of the software vendor) to approximate, often quite closely, those costs.

Carriers, with the "raw" cost information at their fingertips, can make even more sophisticated use of computers for costing purposes. It is possible to identify the incremental costs occasioned by such things as using one terminal instead of another, operating over a particular highway, running or switching rail cars in "cuts" of different numbers of cars, the age of the equipment to be employed, etc., etc. Many carriers (virtually all the big ones) have highly sophisticated costing systems they now operate which allow such computations on an almost instantaneous basis; and even carriers without such "in-house" systems can use commercial software with their own data to achieve very accurate results.

Beyond the process of preparing for negotiations, computers can also lend assistance in other ways to carriers and shippers alike.

Another significant use for computers is to develop, by means of word-processing applications software, "model" or "template" tariff or contractual forms into which the details of a particular such document may simply be interpolated. This can substantially reduce the technical and/or legal problems and delays involved in preparing a tariff or finalizing a contract, and ensure that each tariff or contract will hew basically to the same familiar lines (though the particulars will, of course, differ).

Computers can also prove important tools for both the implementation and administration of contracts—or, for that matter, a shipper's review of the performance of the common carriers serving it. They can be employed to cut bills of lading, freight bills and other shipping documentation; to maintain full and detailed records as to traffic volume, claims, compliance with service-quality standards and other key factors; to transmit, receive, process and even pay bills and/or claims; to monitor timed contractual changes (such as inflation indexing of rates); and to perform countless other contract-related functions.

It is not the place of this book to describe in detail the many computerized applications that may ease the shipper's or carrier's work, or improve its performance, during the negotiations process. But it likewise seems inappropriate to altogether omit mention of the subject, especially in light of the growing use of computers (especially the desktop micro, or "personal," computers) in the business environment. Certainly any negotiator can enhance his results, and make his work far simpler, by the use of computer assistance.

But the computer does not exist—nor does it seem likely to come into existence in the foreseeable future—that can do the fundamental decision-making about transportation rate negotiations. That remains the human manager's role. The computer can provide him with endless reams of information, and analyze that information in a thousand different ways; but it's ultimately up to the manager responsible to make the choices and reach the agreements necessary to establish the business relationship that is the objective of the negotiations process.

7

THE ART OF NEGOTIATING

Negotiating—in business as in any other milieu—is an art, not a science. As such, there is no one "right" way to do it, no single technique or methodology that can be empirically taught or learned.

But as with any other art, there *are* "do's" and "don't's" that can (and should) be taught and learned—certain things that have a proven track record, plus or minus, and can hence serve as guidelines for those who engage in the negotiating process. Some of these are little more than common sense; others are more subtle.

As already discussed, a party's success at the business bargaining table will depend far more on his preparations—his "homework"—than on what he does or fails to do at the table itself. Nevertheless, the tactics of the actual negotiation process remain of importance. Like wars, negotiations must in the final analysis be won (or lost) in the trenches, not on paper; and the best of preparation is not proof against sabotage through negotiating ineptitude.

When, Where and Who

When should negotiations be conducted? If the service involved is continuing in nature, the timing of the negotiations usually won't matter much to either party. It may of course be that other, extraneous factors will lend time an importance here—the shipper may be dissatisfied with his present rates or service and want to finalize new arrangements as soon as possible, or the carrier may be particularly eager to start handling the traffic quickly. But in such cases it will usually behoove the party interested in a speedy result to conceal that interest, since it may be taken by the other (rightly or wrongly) as a sign of bargaining weakness and make a favorable outcome harder to secure.

If, on the other hand, the negotiation deals with a particular movement or group of movements intended to take place at a fixed, future time, it will normally be in the interest of the shipper to start the process as far ahead of that time (within reason) as possible, and in the interest of the carrier to delay it (again, within reason). The shipper who negotiates "under the gun," as it were, has already foreclosed many of his options; he simply hasn't time to talk to very many other carriers if matters don't go his way. Thus, in such circumstances it is obvious that the carrier can afford to take a harder line than might be the case if more time were available.

In the case of railroad contract negotiations, the "when" may take on an additional element of importance because of the potential to apply contract rates *ex post facto* to shipments that moved before the contract was signed. As already noted (*see Chapter 3),* the ICC nominally requires that the negotiation process itself be basically concluded before the rates may be applied to ongoing shipments (even though the agreement may not yet have been put into a signed contract); but this is so clearly unpoliceable as to be little more than a hopeful admonition. *De facto,* there would appear to be little to prevent a shipper from entering into negotiations with a rail carrier at the same time it is giving that carrier traffic to which the contract rate would, after it is decided upon, be retroactively applied.

The actual site of the negotiation is of significance only in two respects. First, it should not be a place where the session will be subject to extraneous distractions. That should rule out negotiations conducted over meals, at cocktail parties and the like. In the first place it's difficult to do serious business while trying to eat. Is that a decimal point or a breadcrumb on the paper? Did this person say "uh-huh" or "uh-uh" when talking with his mouth full? And so forth. Furthermore, alcohol and sound business decision-making mix poorly. And then there are the interruptions of waiters, of other guests at a party, etc.

If the meeting takes place in one of the negotiants' offices, another type of distraction must be avoided—the phone call or walk-in interruption. If day-to-day business is so pressing that a participant cannot afford to shut himself away from it for even an hour or two, he probably shouldn't be sitting at the bargaining table in the first place but should instead delegate that responsibility to others whose time is at less of a minute-by-minute premium; with so much at his doorstep and on his mind, he almost certainly can't give the negotiations the full attention they deserve and need.

Unfortunately, some managers, in a misguided effort to impress others with their own importance, either refuse to give out a "do not disturb" directive or, worse, even set up prearranged interruptions. A few even use the latter as a device they hope will disrupt their opposite number's concentration or impress him with a sense that he is unimportant in the manager's business planning. Unless the visitor has come literally hat in hand—in which case such tactics will probably be superfluous anyway—such displays are more likely to backfire than succeed; the businessperson (whether representing a shipper or a carrier) who is periodically dismissed while the individual with whom he's negotiating takes care of "important" matters will wind up resenting the situation and, if he doesn't break off negotiations altogether, will be far less eager to reach an agreement.

Subject to these qualifications, however, the only importance of the "where" of negotiations (if any) is psychological. Some people feel comfortable only in settings that are familiar, or where they feel in control of their surroundings; if they must instead negotiate in the offices or conference rooms of others—or even at "neutral" sites—they may be considerably less effective at presenting or upholding their positions.* The individual who

*Although existing almost entirely in the minds of those affected, the effects of this phenomenon are quite real, and often significant. For example, any sports fan knows well how the "home field advantage" can affect the

recognizes such traits in his own makeup is obviously best advised to keep negotiations on "home turf"; conversely, the individual who sees evidence of this mentality on the other side should do his best to move negotiations away from that party's offices.

Indeed, when it comes to who should conduct the negotiations, individuals who are highly sensitive to such irrelevancies should probably not be included. What's needed are people who can focus firmly on the business at hand and be full-time advocates for the organizations they represent—and who, if possible, have at least a modicum of negotiating experience.

For the company new to transportation negotiations, or the one that is trying to supplement or replace the individuals who have previously conducted such negotiations in its behalf, this last poses a conundrum that will be familiar to every job-seeker: If what's wanted is experience, how can the inexperienced ever get a chance to obtain it? The answer to this is that, as a normal rule, negotiations should not be conducted by a single individual, but rather by a "team"—a stratagem that offers a number of advantages.

The psychological stresses of the bargaining table can be formidable. Both sides are under at least some pressure to reach an agreement (if an agreement weren't of some degree of importance to both, they wouldn't be sitting down together); and the negotiators for both will normally see the shape of that agreement as carrying implications (real or imagined) about their personal futures within their own organizations. At the same time, both are also faced by offsetting pressures from the other side of the table to give way, to make concessions, to grant benefits—pressures that almost anyone finds at least somewhat difficult to resist when the "no's" must be said while looking the other person in the eye.

---

outcome of athletic contests. To be sure, there will be no cheering crowds present during business negotiations, as there are in sports events; but even without this overt manifestation, many people are acutely aware of the "home" and "away" difference.

Any individual, no matter how experienced, and no matter how seemingly iconoclastic, will find these stresses more difficult to endure alone than in the company of one or more "friendly faces." Even if those who accompany him will have little actual voice in the decision-making process itself, their mere presence will provide the one whose decisions *do* count with a sense of support, both overt and implicit, that may be of considerable help. He may profess to disdain such companionship, even to wish it away, but it is a rare individual indeed who will not secretly be grateful for the presence of others on his side when the going gets tough.

Furthermore, it is likewise a rare individual who can keep personal track of all of the myriad factors that may be of importance in connection with any individual negotiation, especially if he must also be responsible to the negotiating activity itself. In our modern era of specialization, the *bona fide* generalist, even within the context of a particular business activity, is hard to find. And obviously, if information or expertise is needed during the course of the negotiation, it is far better to have it on hand than to have to either base decisions on mere guesses (no matter how "educated") or adjourn matters until it can be procured.

Thus, the inexperienced individual can remedy his deficiency by participating in negotiations in a secondary role, under the leadership of others who *do* have experience. This is by far the most effective sort of training for negotiators—to watch and, over time, to take part in (occupying roles of increasing responsibility), real-world sessions and see how the job is done. Any organization that overlooks this easy opportunity for developing a solid base of negotiating experience within the ranks of its personnel is inviting trouble when the one or two people on its staff who *do* have experience are inopportunely taken ill, go on vacation, move on to other positions (either with the organization or outside it), etc.

(It should be emphasized in this context that the fundamental skills of negotiating are largely independent of the type of agreement that is being negotiated; even though the specifics may differ (perhaps widely), there will be little salient difference between the at-the-table skills and techniques from one economic sector to another. This fact can be of particular importance to shippers, who may lack negotiating experience and expertise in the transportation arena (a not-unlikely circumstance, in view of the relatively recent advent of the deregulation that has made market negotiations so important to the industry) but may be quite accustomed to doing business this way with other suppliers. In such cases the shipper's Traffic (or Transportation, or Distribution, etc.) Department, and the carrier's Marketing/Sales Department, may benefit greatly by working with experienced negotiators from Purchasing or Procurement; and this potential should not be overlooked.)

If experience is totally lacking in an organization that must soon negotiate an important agreement, a partial answer is to set up one or more "trial runs" first—to, that is, negotiate a less-important deal or two before sitting down to discuss the big one. These may even be "make-work" negotiations—negotiations that aren't really vital (the matters involved are fairly small-scale, and would normally be handled through more routine channels), but can nevertheless serve an important purpose in acquainting the organization's personnel with the processes involved without risking serious consequences if mistakes are made.

Finally, the possibility of rehearsals, in which members of the organization take the roles of both sides, is worth considering. Indeed, even experienced negotiators will sometimes sit down to such practice sessions before entering into a major negotiation, just as a lawyer will often rehearse the witnesses he intends to present at a trial before they actually take the stand. Rehearsal (as any theatrical actor can tell you) is no substitute for the real thing; but it can help considerably to prepare a negotiator for the day when he must "play for keeps."

The desirability of placing negotiating responsibility in the hands of a team, rather than a single individual, has already been mentioned. In this context, however, there are several basic rules that should be observed:

- There should always be just one person who is, both actually and apparently (to the other side), in charge. The difficulty of making good, consistent, goal-oriented decisions seems to increase geometrically with the number of individuals who take part in the decision-making process (a camel, it is said, is a horse built by a committee). This is especially so at the negotiating table, where the possibility that commitments agreed to by one individual can be repudiated by his co-negotiator is obviously crippling.

• The in-charge individual should be one empowered by his organization to make decisions that will be binding on it. A negotiated agreement that endures no longer than it takes the negotiators to carry it back to the office is obviously worthless, and both sides have wasted their time working it out; moreover, a shipper or carrier victimized once by such tactics will invariably be twice-shy about returning to the bargaining table with the same party.

• The team should be closely limited in size. Arriving at a negotiating session with a phalanx of "associates" in tow may sometimes intimidate the other side, but more often it will put them on their guard. If sheer weight of numbers ensured success, China would rule the world.

• *Appropos* of the last point, if it is necessary to bring than two or, at most, three individuals to the negotiation—for training purposes (*see above),* to ensure availability of specialized expertise, etc.—the team should be clearly, and visibly (again, to the other side), divided into "primary" and "secondary" tiers. Only the members the first such tier should physically be seated at the conference table itself and/or take a free part in discussions; others should generally be seated separately, and should participate only to the extent specifically requested by their principals.

• If more than one negotiating session is involved, the same team (at least as regards principals) should normally be present every time. To a considerable extent, a negotiation hinges as much on the developing interpersonal relationships among these principals, and these will not develop if the cast of characters is constantly changing. An exception to this rule, however, should be noted where particular individuals on opposite sides of the table form (for good reasons or bad, or even none at all) significant personal dislikes for one another; in such circumstances the personalities are impeding the negotiations process, not helping it, and team membership should be changed if at all possible.*

• The size of the negotiating team should be appropriate to the magnitude of the negotiations being conducted. The parties should not ignore the fact that the other side is aware of the dollar value of man-hours; if one side sends a squadron of half-a-dozen or more individuals to participate in a negotiating session, this accordingly conveys a message that that side regards the session as of great importance; and such a message can have a profound effect on the course of the negotiations themselves. If that is not the message desired to be communicated, the size of the team should be reduced to a level commensurate with what the party *does* want to "say."

Finally, it should be emphasized that successful negotiating "styles" vary widely, and that there is no one "right" way to handle a negotiation. As much as possible, negotiators should be encouraged to present themselves in ways with which they are personally comfortable, rather than trying to ape the manner of others. A good negotiator may be assertive, brassy, even aggressive in his approach—or he may be quiet and self-effacing, or anything in between, without compromising his competence. Negotiating success is measured by results, not attitudes; and the organization should select its negotiators with this firmly in mind.

---

*The same thinking, of course, dictates that individuals who are recognized within their own organizations as possessing especially abrasive personalities not be included as members of negotiating teams in normal circumstances. There may, of course, be cases where this rule should be intentionally violated; in particular, some professional negotiators employ the so-called "good guy/bad guy" tactic, whereby one team principal presents himself to the other side as being especially "tough" while the other plays the role of conciliator. This tactic can be quite effective in some circumstances; but it must be used very carefully, and is not recommended for any but the most experienced of negotiators.

The Initial Proposal(s)

Every major negotiation starts out with a proposal—or with *two* proposals, one from each side of the bargaining table. In many negotiations, however, the initial proposal(s) is (are) not written down. Mostly, that's a mistake.

About the only time a negotiation should not commence with a written proposal is when the initial talks are merely "exploratory," to determine whether the parties have enough commonality of interest to warrant doing serious business together at all. Once this stage is past—once each party has independently decided that it would be fruitful to consider a business relationship with the other—matters should start being reduced to writing at every stage in the negotiation.

These written drafts provide, in effect, an agenda for each negotiating session. They also do much to reduce the need for an excessive number of such sessions, each of which, of course, costs money and occupies valuable managerial time. In both senses they expedite the process considerably, and hasten the day when the parties can stop merely *talking* about doing business with one another and commence actually doing it.

In transportation negotiations the original proposal will normally be drafted by the shipper, he being the party with the clearest knowledge of the traffic involved, the service requirements, etc. But the carrier should not come to the first meeting empty-handed; after all, he has his own proprietorial information (as to operating costs, etc.) which is unknown to the shipper, and in most instances will have established tariff rates which provide the starting point from his perspective.

Obviously, neither party should incorporate its "best shot" into its initial proposal—and it can be sure the other side won't, either. Each side's initial proposal really amounts to a written version of its "wish list," rather than what it truly expects the final agreement to embody—and the other side will certainly look at it that way.

Nevertheless, there is a distinct tactical advantage to presenting the initial proposal in clear written form, approximating as closely as possible the form and language of carrier tariffs (or, if contract carriage is intended, a written contract). This conveys the message, in effect, "Here's my deal—and I'm ready to go forward right now!" Such a ploy tends to discourage the other side from bargaining too intensively for non-essential concessions, since each separate provision, already down on paper, must be dealt with specifically and individually.

Since both parties will (in this recommended scenario) be seeing one another's proposals for the first time at the initial meeting, it would be unrealistic to expect a final agreement out of that meeting. Indeed, should such an agreement be forthcoming it is probably a sign that one of the parties has asked for too little—for less than it might, by more aggressive bargaining, have been able to achieve.

With at least the shipper's, and possibly also the carrier's, proposal(s) on the table, the negotiations may now commence. At this juncture each party will have different tactical objectives, each trying to ascertain the other's "bottom line" without revealing too much of its own. When experienced negotiators are sitting on both sides of the table, there is apt to be substantial time taken up in determining the protocol of the negotiations, each trying to put the other in a position where it will have to commit itself first.

Specifically, the shipper should be looking for information concerning minimum acceptable rate and charge levels, operational capabilities, etc., of the carrier. Meantime, the carrier should seek to identify the precise quanta of service sought by the shipper, the actual anticipated volume of the traffic, and other details. Most of the first meeting, in a serious and well-handled negotiation, will probably be dedicated to this "feeling-out" process, with both sides reluctant to make major concessions or commitments.

Such a meeting will commonly end with both sides having a much clearer idea of (1) what the other ideally wants (*i.e.,* its initial proposal), and (2) what it is willing to settle for. There will have been no substantive agreement at this stage; rather, each will carry back the information it has gleaned to its own offices, where fully detailed analyses can be made.

(It is, to be sure, possible that negotiations will have progressed further than this. Many individual are intrinsically wedded to the concept of rapid decision-making, and indeed take pride in how rapidly they are able to decide matters. Even the fastest decision-maker, however, will usually benefit from an opportunity to "sleep on it" before committing himself; and for that reason it is normally unwise to offer any substantial commitments at the initial meeting, after having only a few moments to review the other side's desires.)

At this stage both parties must make the critical decision: Is there a basis for an on-going relationship, or is there not? Many negotiations break off after the first meeting, relatively few after that; the decision even to have further meetings is, in the widely accepted conventions of negotiation, a broad commitment to pursue the matter through to conclusion and agreement.

The point is that each now knows enough about the other to make a rational decision in this area. It is fruitless to spend hours or days seeking to break what is obviously, from the start, an irreconcilable deadlock. Thus, the "go" or "no-go" decision is ready to be—and should be—made at this point.

Assuming both parties decide they are not so far apart that an agreement appears out of reach, each should now be in a position to formulate what it anticipates would be a reasonably equitable agreement. And that is precisely what they should do, again putting their proposals down on paper (in the rough form of a tariff, rate circular, contract, etc.).

Now there are two options; the one that should be exercised depends on how close each side perceives the two have come to a final agreement. If agreement appears fairly close, the best course of action may be for each to submit its proposal (by mail, hand delivery, Telex, etc.) to the other side for review without a subsequent meeting. If, on the other hand, there still seems to be a fair negotiating distance to traverse, it may be better to schedule a second meeting at which the proposals will be both presented and discussed. Each side may either make its own determination in this regard, or there may be an agreement reached at the close of the first meeting.

In either case the second meeting should be aimed, insofar as possible, at arriving at what will be the terms of the final agreement. Bear in mind that most transportation negotiations are probably not of the "make or break" variety for either party; such negotiations do not usually carry the same overtones as, for example, labor negotiations, or major supplier-vendor negotiations, etc.* It is, accordingly, desirable to limit the amount of time and energy dedicated to them.

As a general rule, if agreement cannot be reached in two or (at most) three meetings, it will probably not be worth the extra effort to pursue the matter further—on the general premise that disagreements that drag the negotiations process out further than this will usually be found to represent an unresolvable impasse. The only major exception involves cases where the negotiations have taken an unhappy turn due to largely extraneous factors—personal conflicts among the negotiators, internecine disputes within the organizational structure of one of the parties, etc.—which can be dealt with or eliminated from future discussions; or where external factors cause a substantial change in the facts and circumstances that underlie the area(s) of key disagreement.

The Conduct of Negotiations

As already discussed, there is no one "right" style or approach to the negotiation process; individuals do best to take whatever tack is most personally comfortable to them, rather than trying to comport themselves according to some arbitrarily defined "ideal." There are, however, a number of broad guidelines that may be offered as to actions and behaviors which will generally contribute to a successful result, and those which will usually prove counter-productive:

*To be sure, there are exceptions to this, as where "captive" traffic is involved, or where a carrier has committed itself so deeply to a shipper than it cannot survive if that shipper's traffic is withdrawn—and in such cases considerably more time and energy may have to be dedicated to the negotiations process.

• A strong effort should be made by the negotiators to develop a good rapport with their counterparts on the other side of the table. It may seem odd that, in the supposedly impersonal world of business, personalities still play a key role in many relationships; but that impersonal world is in fact peopled by human beings, and human beings are a social species; they prefer to do business with those with whom they are on friendly terms, and will often stretch a point or two in pursuit of that desire.

• By the same token, rancor, overaggressiveness, hostility and the like are almost *never* useful attitudes to display at the negotiating table. If they in fact represent the true feelings of a negotiator, he should either bend serious efforts toward disguising them or, if he cannot do so successfully, excuse himself. Even if the other side behaves badly, that behavior should not be returned in kind; a negotiation that becomes a battlefield will leave neither side the victor. Courtesy and good manners were invented to make it possible for people to associate profitably with one another; their absence will not serve any useful purpose.

• Courtesy and friendliness should not, however, be carried to extremes. It is said that in some societies, if one expresses great admiration for the possession of another, that other is expected to offer it as a gift. This is perhaps valid as a rule of social intercourse for individuals brought up in the niceties of such conventions; but anything remotely approaching it in a business negotiation would obviously be disastrous for the party taking such an attitude.

• Carrying this a step further, negotiating concessions should rarely, if ever, be granted without consideration. The business negotiation is based on the barter system—tit for tat, give and take. To give way on a point without receiving in exchange a concession of at least roughly equivalent value is to reveal a basic weakness in one's position, which one can then expect the other side (if it's at all alert) to exploit fairly ruthlessly.

• Clear and precise communication is essential to any negotiation. This is a basic reason why it's desirable to put proposals in writing; the spoken word is fleeting, and can inadvertently (or intentionally) deceive. At each stage—at the point of each new level of agreement—the parties should reduce their understandings to writing, and should give one another clear "feedback" about what has actually been agreed to.

• Many behaviors may be appropriate to the bargaining table; dishonesty is not among them. Outright lying, or even knowingly misleading "truths," should be avoided at all costs. This does not, of course, mean the parties are obliged to reveal their closest secrets to one another; if information is proprietary it may certainly be held as confidential, and requests for its disclosure be turned away on that basis. But the outright untruth can only be destructive; even if it is not uncovered during the negotiations themselves, it is apt to surface later on and sour the business relationship the negotiations have engendered.

• A bargain is a bargain, and a deal is a deal; once made, concessions and agreements should be strictly adhered to, even if second thoughts reveal them to have been unwise. If the change of heart occurs at the bargaining table, the party that does so reveals itself as too inept even to know its own mind; if it takes place after the negotiating session, following consultations back at the office, the message is conveyed that the organization has no faith in those it has itself chosen to represent its interests. And in either case, the other side has good cause to wonder whether, even if an agreement is ultimately reached, it can be considered truly binding. The basic rule of thumb is that, if one doesn't mean it and/or isn't willing to stick to it, he shouldn't say it to start with; but if one does speak too hastily, at least he should have the integrity to live with his mistake.*

• Finally, even if the negotiations fail to bear fruit the session should not be allowed to end in acrimony if this can be

*Obviously there are circumstances under which this rule must be disregarded. If the mistake is truly egregious, to the point where it will be seriously debilitating if not corrected, sticking to the letter of one's agreement would be foolhardy. But if such a situation arises, the party that must renege should explain the circumstances as fully as possible to the other; and even then it will have to recognize that its posture for future negotiations has been seriously compromised.

avoided. The fact that the other side may not be willing to yield to one's own position in a particular instance does not necessarily make it unreasonable; it may well be that the interests of the two sides simply don't adequately coincide to make a business relationship between them mutually profitable—*this* time. But it may also be that the selfsame parties would ideally complement one another in some future, perhaps now-unforeseen, circumstances. The wisest course, if negotiations fail, is to leave the door at least ajar for possible future business dealings, rather than to vent one's own frustrations at the expense of the opportunity to perhaps strike a mutually desirable agreement tomorrow.

'Boulwarism'

Most of the foregoing has been based on the premise that the parties are meeting at the bargaining table as, at least ostensibly, relative equals—that is, that they are prepared to make concessions of their own as well as to seek them from across the table. In one approach to the negotiation process, however, this will not be the case.

About half a century ago, when the U.S. labor union movement was still struggling against oppressive anti-labor laws and widespread public opprobrium, certain companies made use of a labor bargaining tactic that became known as "Boulwarism."* The employer would develop a proposal it deemed "fair" to both labor and management, present it to the union (usually accompanied by much fanfare designed to enlist public support on its side)—and then refuse to budge on any of the key points of that proposal. The union could accept the proposal, or could strike (in those days a risky proposition indeed from labor's standpoint); but it had no meaningful third alternative to this kind of "take-it-or-leave-it" offer.

---

*This tactic was most prominently espoused by, and is named after, one Lemuel Boulware, a General Electric Corp. official of that era.

Boulwarism has long been outlawed as a labor negotiating tactic by the National Labor Relations Board and the courts; today such intransigence is deemed a "refusal to bargain," for which various legal penalties are prescribed. But it survives in other areas where "public policy" does not so deeply concern itself with the ways in which negotiations are conducted, and has found occasional use in the deregulated transportation marketplace.

For Boulwarism to prove effective, at least one of two conditions must prevail. Either the proposal must truly be what it purports to be—that is, genuinely reasonable for both sides—or the party to whom it is presented must have no realistic option but to accept it. Since it is extraordinarily unlikely that both parties will see the concept of "reasonableness" in exactly the same way, it is in the second circumstance that Boulwarism normally flourishes.

In particular, shortly after motor carrier deregulation took place (at the Federal level) in 1980, certain major shippers adopted the practice of "publishing" their own truck "tariffs." These documents were handed out on request to all motor carriers interested in serving such shippers, with the further advice that any motor carrier need only write its own name in the blank left for that purpose and file the document with regulatory authorities in order to participate in these shippers' traffic. Those that did so (out of fear of the deregulation-engendered competition they faced) found themselves, of course, saddled with rate levels that were barely (if at all) profitable; but those that did not were generally cut off from the traffic in question.

In some instances shippers continue (as of this writing) to at least nominally espouse this "negotiating" approach. Most have found, however, that they get better results by dealing more flexibly with their carriers; and even those who give lip service to Boulwarism find that, in practice, they can rarely get adequate levels of service by means of this kind of rigidity.

More problematically, some railroads take a more or less similar attitude toward captive traffic. Feeling that they "own" the traffic in question, they largely decline to enter into meaningful talks with their shipper(s), setting rate and service levels more or less as they choose. In the long term, again, such contempt for the negotiating process is likely to prove self-defeating, one way or another; meantime, shippers confronted with this tactic can deal with it by adopting one of the various approaches for dealing with such difficult captive-market situations (*see Chapter 5).*

Finalizing the Agreement

Finally, the negotiated agreement must, of course, be reduced to written form—incorporated in the provisions of a tariff, rate circular, contract, etc. Here the important thing is that the writing accurately reflect, in legally binding language, what was agreed. To this end it will usually be desirable to draw in those expert in the written format involved—in the case of common carriage, experienced tariff writers and readers.

Accurate and comprehensive drafting of the documentation is vital. The entire negotiation is rendered meaningless if the parties wind up seeking to subvert it through such devices as obscure tariff or contractual language or interpretations. The substance of the agreement, and not trickeries or technicalities of form, should rule the relationship.

Note that this is emphatically *not* to say that, where common carriage is involved, shippers should abandon efforts to hold carriers to the letter of *non*-negotiated tariff provisions. The carrier is, after all, unilaterally responsible for preparation and publication of the tariff; and the shipper has every entitlement (both legal and moral) to construe that tariff in the manner most favorable to it.

But such an attitude is inappropriate to tariff provisions that result from bilateral negotiations, are aimed at reflecting the outcome of those negotiations, and hence are in fact *not* truly the unilateral creation of the carrier. Here, even should the carrier be careless or unwise enough to lay itself open to applications, interpretations, etc., more favorable to the shipper than contemplated by their agreement, the shipper has the ethical responsibility of living up to its end of their bargain. Of course, the same holds true for the carrier.

Because of the legally binding nature of transportation tariffs, this may be problematic if the tariff is poorly worded or—especially as to rate levels, where a simple typographic error can produce grotesquely wrong results—carelessly produced. The simplest and most straightforward answer is for both parties to review the final tariff before it is actually published. Additionally, it may also be desirable for both to sign a separate paper (letter form will do) declaring the manner in which they intend to interpret and apply that tariff; in light of new freedom available to carriers to interpret even ICC-published tariffs as they see fit (*see Chapter 3),* this should avoid any legal problems satisfactorily.

At the same time, none of this relieves the carrier of its responsibility to ensure that the tariff does, accurately and completely, reflect what has been negotiated. The case books of the ICC and the courts are replete with records of problems having arisen due to careless or inept drafting of tariffs; and the hard truth is that the vast majority of such problems are ultimately resolved in the shipper's favor. It is foolish to open the door to such problems when a reasonable amount of care and skill applied to the original drafting of the tariff language could eliminate the potential for them.

Finally, it should be emphasized that no agreement should *ever* be deemed truly final until it has been put into writing. Once again, the ephemeral quality of spoken words, no matter how earnestly meant at the time they're said, must be noted. This is especially true where the law requires publication of such shipper-carrier agreements in tariff form, and insists that published tariffs be the sole basis for the provision of transportation services and the assessment of rates therefor. A great many shippers, during the early years of deregulation, were caught in the trap of accepting non-tariff "discounts" from carriers who subsequing the early years of deregulation, were caught in the trap of accepting non-tariff "discounts" from carriers who subsequently (either themselves or by proxy) dunned them, usually successfully, for "undercharges" based on the much higher rates that in fact *were* in their (the carriers') published tariffs.

Two 'Winners,' No 'Losers'

Perhaps the most important single consideration in any form of business negotiation is that both sides must always endeavor, at least to some degree, to examine any agreement they reach from the other's perspective. It is much the same as with the compensation paid to employees; either one must pay what is largely regarded as a living wage or one must accept, in consequence of one's failure to do so, a very high degree of employee turnover. Shippers who insist on rates viewed by carriers as noncompensatory will have to accept a high drop-out rate among the carriers serving them—either through defection or (probably a lot worse) through insolvency and bankruptcy. And, to turn things around, carriers who push excessively hard for unusually high rate levels, even on captive traffic, are apt to pay for it later.

This is not, of course, to suggest that it is incumbent on shippers to look out for the economic welfare of the carriers with which they do business, or that carriers should coddle their customers. Not being—at least with respect to their commercial dealings with one another—eleemosynary institutions, neither has any such obligation. At the same time, most shippers want to continue doing business with the same carriers over extended periods of time, since this saves the administrative and operational problems associated with having to "break in" new suppliers. And most carriers would prefer to establish dependable "stables" of customers on whom they can count for future business, in order to provides themselves with a firm marketing base on which they can build. Moreover, shippers want good service, which rock-bottom freight charges usually won't buy them; and no carrier wants a blizzard of complaints, claims, etc., from unhappy shippers.

In these circumstances, both should avoid situations in which the party on the other side of the bargaining table comes away perceiving itself as the "loser" of the negotiation. Ideally, both should emerge thinking of themselves as "winners"; of this will the most satisfactory and lasting of business relationships be created. But at worst, no matter how it sees the negotiation result itself, each side should do its best to ensure that the other does not feel unduly defeated, and therefore cultivates resentment. In any really fruitful business relationship the parties must actively desire to do business with one another; any party that does not will, purposely or subconsciously, be working from that point forward to sabotage the relationship—and will, in all probability, sooner or later succeed.

Appendix
Negotiations in Transportation*

Scene: Conference table, with four empty chairs around it. Business-type phone console at one end of the table.

CHARLES *(voiceover):* Memorandum. From Charles Parsons, Traffic Manager, XMZ Manufacturing Corporation. To Denise Wachower, Vice President - Distribution. Denise, I'm having a problem with truck service to our distribution centers in Austin and Dallas, Texas, and Oklahoma City. It's tough to get carriers to go down there at all; we get three of four times as many refusals on this traffic as to anywhere else we serve. And even when we can get equipment to load, service tends to be sloppy and the rates are extremely high—sometimes I have to pay as much as $1.50 a mile.

I'd like to recommend a different approach here—offering most or all of the traffic to a single carrier, either on a common or contract basis. It might be that, with a substantial assured volume they could count on, some carrier would give us more reliable service and better rates. We might even lump this Texas-Oklahoma traffic with some other movements the carrier would find more desirable, as a kind of sweetener.

I haven't discussed this with any carriers yet, of course; I wanted to get clearance from you on the idea first, and also it might be better if the contact came from you. Hope you can help me out with this problem; it's really getting to be a headache.

DENISE *(voiceover):* From Denise Wachower, Vice President - Distribution, XMZ Manufacturing Corporation. To Melvin Langden, Executive Vice President, Victory Trucking Company. Dear Mr. Langden: As I'm sure you're aware, we have substantial movements of product outbound from our Chicago facility to various destinations. In the past we have been distributing this traffic among a number of common motor carriers that have qualified on our eligibility list, including Victory Trucking Company.

---

*This script was used as the basis for a model transportation negotiating videotape produced by the author under the aegis of the Academy of Advanced Traffic, a subsidiary of the publisher of this book. Copies of the videotape are available for rent or purchase through either the publisher or the Academy (819 Meetinghouse Rd., Cinnaminson, NJ 08077, telephone (609) 786-9113).

At this time we are considering a change in this operation, whereby one carrier would be identified as prime supplier for movements to grouped destinations, and would handle the bulk of our traffic to a given destination grouping.

I would be interested in discussing with you the possibility of Victory Trucking Company becoming our prime supplier on movements from Chicago to destinations located in what we have identified as Destination Grouping Seven, consisting of the states of Arkansas, Louisiana, Missouri, Oklahoma and Texas.

Please let me know if you are interested in pursuing this opportunity. I would be pleased to meet with you at your convenience to provide you with further information concerning the specifics of this traffic, and to receive your proposals concerning service and rate levels.

Yours sincerely, et cetera.

JOHN *(voiceover):* Memorandum. From John Prudella, Marketing Manager, Victory Trucking Corporation. To Melvin Langden, Executive Vice President. Reference the XMZ letter you asked me to look at: Mel, XMZ has given us a problem because of the way they've grouped their traffic.

If we don't go for this proposal it looks like we'll lose the New Orleans and St. Louis business we're currently handling for XMZ, or most of it; and that traffic is proving extremely profitable because of the backhaul contracts we have set up from those points. But the price of keeping that traffic is apparently taking their movements to Arkansas, Texas and Oklahoma as well.

Arkansas is no trouble; they have only a small volume into Little Rock, and I have some backhaul prospects in that general area. Texas and Oklahoma, though, are another matter; they have heavy volume to Austin, Dallas and Oklahoma City, and you know how lousy the backhaul situation is from there.

They've boxed us pretty good, and I'm fairly sure they know it. I suppose we could cover the New Orleans and St. Louis fronthauls with other shippers, but it would be a scramble and I'd rather not have to do it. I do have one prospect in the Dallas area, which would cover our backhauls out of both there and Oklahoma City; I've been hesitating on this one because the revenue is awfully low, but if we could get a decent fronthaul rate we could still get some contribution to overhead out of the round trips. That leaves Austin, and all I can say is that we need to keep the rate as high as possible and even then we may have to eat some loss there; but it looks like that's the price of hanging onto New Orleans and St. Louis.

I think you and I should get together with the XMZ people and see what we can work out on this.

On screen, camera shifts to door, showing people entering. First come CHARLES and DENISE, who take their positions on one side of the table; then, with enough pause to clearly differentiate between the two groups, come MEL and JOHN, who greet CHARLES and DENISE and then take their places on the other side. Appropriate removal of papers from briefcase, paper shuffling, etc. This all takes place during the final voiceover:

MEL *(voiceover):* From Melvin Langden, Executive Vice President, Victory Trucking Company. To Denise Wachower, Vice President - Distribution, XMZ Manufacturing Corporation. Dear Ms. Wachower. I read with interest your letter concerning realignment of your distribution plans, and have reviewed with our marketing and operations professionals your suggestion that Victory be your prime supplier for transportation service to your Distribution Grouping Seven.

We would indeed be interested in discussing this prospect with you, and would suggest a meeting to pursue this matter. At that time I would hope to be able to get fuller information concerning the traffic volumes and destinations involved, and will be prepared to respond with a proposal that I hope will you will find attractive.

My schedule for the next week or two is reasonable clear, so I'll leave it up to you to suggest an agreeable meeting date. I look forward to seeing you at that time.

Action shifts to on-screen.

DENISE: Gentlemen, I'd like to start by briefing you on our Distribution Grouping Seven. Mr. Parsons has the details. Charles?

CHARLES: Yes. In terms of commodity, the traffic is plastic products intended for various uses. These products have an average density of 10 pounds per cubic foot, with a range between a low of seven pounds and a high of 15. In terms of class rating, the range is between class 77-1/2 and class 125, although only about 10% of the mix is rated at over class 92-1/2. I have the detailed breakdown here; overall, and taking into consideration of course that there will be variances shipment by shipment, the mix will be about the same for each of the destinations. *(Passes sheet of paper to each man.)*

Now, the destinations: In the area we've identified as Distribution Grouping Seven we have six distribution centers: Austin and Dallas, Texas; Oklahoma City; Little Rock, Arkansas; St. Louis; and New Orleans. I believe you've been handling considerable traffic to St. Louis and New Orleans, but relatively little to the others.

The largest volumes go to Austin, Oklahoma City, St. Louis and New Orleans; to each of these we average five to six shipments a week. Dallas takes somewhat lower volume, perhaps three to four shipments a week; and Little Rock is a subordinate location with only one or two loads per week at most. All the traffic is truckload.

All right, in a nutshell that's it. Again, I have the figures written down for you. *(Passes second sheet of paper to each man.)*

JOHN: OK, this is fairly helpful, but I think we'll need a bit more. For example, what about packaging? The loads we've been handling to St. Louis and New Orleans have been in cartons shrink-wrapped onto pallets, usually about ten pallets to a truckload. Is the other traffic the same?

CHARLES: Most of it is; but we don't have fork equipment at either Austin or Little Rock, so movements there have to be floor loads.

MEL: Does that mean we'd do the loading and unloading on that traffic?

CHARLES: Yes, that would be expected; we don't have the personnel available to handle floor loading or unloading.

MEL: I see. . . .

JOHN: Aside from these movements, I presume we're talking about pallet return.

CHARLES: Yes, that would be included, just as it is with our present movements.

JOHN: Any protective service requirements?

CHARLES: No, the stuff is pretty temperature-insensitive.

MEL: John, do you have anything else? *(JOHN shakes his head no.)* All right, then, we've worked up some cost figures back in the office, and figure we can give you some pretty favorable rates on these movements. Now, rather than get into class distinctions on this traffic I'm going to quote you on a freight-all-kinds basis, point-to-point.

Now: Our proposal is to offer you $1.10 to St. Louis, $2.59 to Little Rock, $3.52 to New Orleans, $3.56 to Oklahoma City, $4.19 to Dallas, and $5.36 to Austin. These will all be based on a 30,000-pound minimum, which seems most appropriate to your traffic. We've figured these rates on a class 85 basis, which I think is about the average of your traffic; and each of them is at least 7% below our present level, and in some cases we've incorporated reductions of nearly 10%. We're able to offer you these reductions on the premise that, as your primary carrier to your Distribution Grouping Seven, we'll be able to count on steady and consistent volumes of traffic to these points. I have the rate information written down here. *(He passes around sheets of paper.)*

DENISE: I wonder if I could ask you to calculate these rates on a cents-per-mile basis? That should give us more of a common denominator for comparison purposes.

MEL *(hesitant):* Well, yes, I suppose we could do that. John, do you mind calculating it out? *(JOHN produces a calculator which he starts using.)*

You have to understand that our costs vary significantly from one route to another; so we can't base a quotation on the same cents-per-mile revenue for these very different destinations. *(Looks around expectantly, but DENISE and CHARLES make no response.)* In particular, we incur especially high costs serving points in the Texas-Oklahoma region, and our rate levels naturally have to reflect that.

John, do you have the figures?

JOHN *(finishing up a final calculation and jotting down the result):* Yes, all right, to St. Louis and New Orleans both, this works out to $1.16 a mile. Little Rock is $1.23 a mile. Now, as Mel mentioned, our costs to the other destinations are appreciably higher, but I think we're still extremely competitive. The Dallas and Oklahoma City movements work out to $1.37 a mile, and the movements to Austin would be $1.45.

DENISE: That's based on your minimum, 30,000 pounds?

JOHN: Yes.

DENISE: That's a fairly high minimum for us. Most of our shipments don't weight out quite that high; I think our average is—Charles?

CHARLES: It's about 27-28,000.

DENISE: So in most cases we're going to have to pay deficit weight.

MEL: We can reduce the minimum if you like; but the rates we've quoted you are based on 30,000, and reflect our costs here. The only way we could really reduce it would be to increase the actual rates to compensate.

DENISE: Yes, I understand, but suppose we make it a flat truckload rate. That way, where we do happen to have a shipment that goes over 30,000, we at least don't have to pay a penalty in the form of higher charges.

JOHN: Well. . . .

DENISE: If the revenues at 30,000 cover your costs, wouldn't higher revenues from the few heavier shipments really be a kind of windfall? The additional weight really doesn't affect your costs, and I think if we're going to pay deficit weight on most of our shipments you ought to be willing to meet us halfway and waive weight charges over 30,000 pounds.

MEL: All right; we really hadn't considered the question of loads over 30,000 when we put our numbers together, and I guess we can restructure these rates to a flat per-truckload basis without regard to weight.

One more thing: When we figured these rates, we assumed all the movements would be palletized. With the higher costs involved in our loading and unloading the freight to Austin and Little Rock, we'll have to add a 10 cents a hundredweight charge on each end of these movements—a total of 20 cents if the shipments are handled as floor loads at both ends, or 10 cents if you load palletized or on slipsheets and only ask us to unload.

CHARLES: Um. . . I'm not. . . Correct me if I'm wrong, but doesn't the present tariff rate include carrier loading and unloading?

JOHN: Well, as we pointed out, your current shipments, the ones we've been handling for you, have all been palletized and mechanically loaded and unloaded by your people.

CHARLES: Yes, but that's not a tariff condition, is it? I mean, the same rates are good on carrier-loaded shipments, aren't they?

JOHN: Well, technically speaking they are, but as I said—

DENISE *(interrupting):* I think Charles has an excellent point here. You say you're offering discounts of 7 to 10% off the existing tariff rate because you'll be handling larger and assured volumes, but then you restrict that discount to a further condition that's not in your tariff now—that we ship palletized and handle all loading and unloading ourselves. There's obviously a substantial cost savings to you here. You've put a value of 10 cents a hundred on that, and I'll accept your valuation; but if we plug that additional factor into the equation, doesn't it just about offset the discount you're proposing?

MEL: Well, I haven't actually done the math on that.

DENISE: No, but aren't I essentially right? And if I am, what's happened to the discount based on the additional, assured volume?

JOHN: Well, but you're not taking a real-world view here. I mean, you actually don't take advantage of the tariff provision permitting carrier loading and unloading, so there's no real-world value in that for you.

CHARLES: That's my fault, I guess. *(Turns to speak to DENISE.)* I should have asked for a loading allowance a long time back, and I just overlooked it.

DENISE *(gives CHARLES a rueful shrug; then, to MEL and JOHN):* Well, one way or another it's obvious that this situation shouldn't be allowed to continue. So, as matters stand, you're offering us nothing much better than the actual tariff rates we're paying now. I think with the volume we have in mind we'll really need something a bit better in order to justify naming you as our prime Distribution Grouping Seven carrier.

MEL: I take your point, and from your side of the table I think it has a lot of merit. But I'd like you to look at it from our perspective, too: We've been talking about "assured" traffic volume in a general way, but actually there really is no "assurance" at this point, is there?

JOHN: Yes, what happens if we give you a volume discount and then a couple of weeks or months from now one of our competitors goes down another notch and the bulk of the traffic shifts away? Essentially we'll have traded off significant revenue for nothing but a hope.

DENISE: I see what you're saying. All right, suppose we incorporate some kind of volume commitment?

MEL: What sort of commitment did you have in mind?

DENISE: I'd suppose some kind of periodic volume minimum. If we meet it, we're entitled to a discount greater than you've proposed so far; if we don't, we pay the rate levels you've just proposed.

MEL: I think that would be acceptable, although it would seem to me more appropriate to go back to our existing tariff if you fail to meet the volume standard.

CHARLES: OK, but with an additional provision for a loading allowance on palletized movements if we wind up going back to the tariff. You've set the number yourself—10 cents a hundred at either end of the movement.

MEL: All right. Now, I gather that we're talking about a volume commitment more or less equivalent to the volumes you described—say six loads a week to the four major centers, four to Dallas and two to Little Rock.

CHARLES and DENISE *(roughly together):* No—

CHARLES *(to DENISE):* Sorry.

DENISE: Mr. Langden, please understand that although we've given you this volume information in good faith and it represents our best estimate, that's still a long way from a commitment that traffic will *actually* equal or exceed those levels. We have a relatively stable market, but, like anyone else, we're subject to unexpected shifts in the economy, changes in industrial and consumer buying patterns, changes in the market for our raw materials, and so on. I think we'll really need more flexibility than you're suggesting.

MEL: Well, what did you have in mind, then? Obviously we can only base discounts on the minimum-volume guarantee, and the lower the guarantee. . . .

CHARLES: I don't think we need to work it quite that way. As we've said, we're hoping you'll be our primary carrier for the Distribution Grouping Seven market area. In those circumstances we'd naturally expect to tender you the bulk of our traffic in any case. Why not simply express our volume commitment on that basis—guaranteeing Victory, for example, 80% of our total traffic to these destinations?

DENISE: In effect, we're asking you to share both the risks and the rewards of our market. If our business falls off, so does our traffic and you share our bad luck. But if our market picks up, again, so does the traffic, and you share our success there.

MEL *(slowly):* I think that sort of arrangement would be acceptable. I agree with your view that your market is fairly stable, so the risk shouldn't be excessive here. But I'd be a lot happier with a 90% participation guarantee than with 80%.

CHARLES: I think we can do that, provided you can make equipment available on reasonable notice. Obviously, if we have to call on other carriers because you can't provide the service, that tonnage is going to have to be taken out of consideration.

JOHN: What do you call reasonable notice?

CHARLES: The worst we'll give you is overnight—notification one afternoon for pickup the next morning. And in most cases our loads will move on schedules developed two or three days ahead.

JOHN: I think we can live with that pretty well. But I do have one point I'd like to add about the volume question: I think the 90% should apply separately for each destination. It could be very awkward for us if one route suddenly fell apart on us because you took the entire 10% overflow out of that one.

CHARLES: How about 90% overall, 80% to any given destination?

JOHN: All right, I can go for that.

MEL: Well, with this level of volume commitment, I don't have any trouble making some reductions here. In fact, I think I can manage a full 5% across the board, based on the proposal I gave you a few minutes ago. *(JOHN is busy with a pocket calculator.)*

DENISE: Now what does that bring us to?

JOHN: Again talking in cents per mile, for New Orleans and St. Louis we're offering $1.10 a mile, which, as you know, is about rock-bottom for our industry. For Little Rock, $1.17; for Dallas and Okie

City, $1.30; and for Austin, $1.38.

MEL: And I really don't see how you can do much better anywhere; as John said, these are really favorable rates.

DENISE: Well, I agree we've made considerable progress. But I think there are still a few issues we need to explore here.

CHARLES: Let me start with the question of delivery scheduling. This is fairly important to us; we're trying to keep inventory down to pretty bare-bones levels, so we need both quick and, even more important, *reliable* service here. Now, the scheduling I'm about to suggest presumes loading is handled early morning and is complete and the vehicle has been released by no later than noon.

We run two shifts in St. Louis, so we can accept delivery well into the evening; we'd like to count on same-day service there. For Little Rock we'd want first-afternoon delivery; for Oklahoma City, second-morning; and for the other points, second-afternoon.

JOHN: Are all these facilities two-shift like St. Louis?

CHARLES: Little Rock and Dallas aren't; the others are.

JOHN: Well, we can handle the same-day St. Louis service. And second-afternoon will work out for New Orleans. But the others. . . you're pushing us pretty hard here.

MEL: I think we'd need to put on tandem equipment to meet the Little Rock and Austin schedules—and that would knock our costs all out of line, based on these rates.

JOHN: Suppose we try easing back just a little. You've made your point, that time is really important to you; now let me see what I can realistically do. *(Uses the calculator briefly.)* OK, suppose we try second-morning to Little Rock; since you're not open late there, I can't count on the afternoon delivery. Second-morning to Oklahoma City just won't work—that's an 800-mile haul, and you're asking me to make it in a day and a half. Second afternoon there, if you want it; otherwise, third morning. Dallas we can make second afternoon all right; but I think we'd better figure third morning to Austin unless you want delivery awful late—say, after 6:00 p.m.

CHARLES: After six is OK for Austin; and I *would* like second afternoon for Oklahoma City. With those changes, I think the schedule is acceptable. We'll plan to give you delivery appointments when the truck is released, within the limits we've just agreed on.

DENISE: In fact, I'd really like to work out some kind of guarantee about your making these appointments—perhaps a penalty if you're late.

MEL *(laughs):* Ms. Wachower, I really admire your, uh, enthusiasm, but I think a penalty is fairly unrealistic with the discounts we're offering here. You're giving us no leeway at all to allow for the unexpected, the unforeseen—

JOHN *(picking up the theme):* Equipment breakdowns, accidents, driver illnesses, traffic tieups, weather—

DENISE *(interrupting):* Then you can't really meet the schedules we've been discussing?

JOHN: No, now, certainly we can meet those schedules under normal circumstances. But I'm just saying—well, you said it yourself earlier when we were talking about traffic volume; an expectation, even in the best of faith, is a long way from a commitment.

CHARLES: OK, suppose we approach it the same way, then, using percentages. You say you can meet these schedules under normal circumstances, and since we're talking about some pretty sizeable volume here we should be able to deal with *ab*normal circumstances more or less statistically. Suppose we say you guarantee 95% on-time, and any loads below that level move without freight charges?

MEL: Suppose we say 90% and a 20% discount for lateness beyond that? *(CHARLES starts to speak, but MEL stops him with a gesture.)* Look, you're getting more or less something for nothing here; normally there aren't any guarantees at all, and you have to accept reasonable dispatch. I really don't think I can go any better than 90 and 20 without having to reflect it in the rates.

DENISE: Well, I think we can settle on that for now, although we're obviously going to keep an eye on delivery timing as we go along.

JOHN: Frankly, I think that's really the answer to any question of service quality. Look, all this talk about on-time percentages and penalties is really meaningless in light of the fact that, as the shipper, you can always simply shift to competing carriers if our service doesn't measure up to your standards. Now, I really don't think there are going to be any problems; at Victory Trucking we pride ourselves on service quality—

CHARLES *(interrupting firmly, though not harshly):* I understand your feelings; and bear in mind that, if we weren't fairly well persuaded that you can get the job done for us, we wouldn't be talking with you to start with. But we've just finished agreeing on a volume commitment that's going to penalize us substantially if we do shift to other carriers, and in these circumstances I think we really need some form of service guarantees beyond the norm.

MEL: You know, as we've been talking here it occurs to me that what we're really discussing is a contract, rather than common carriage.

DENISE: I agree; I was about to bring up the same point. Since we're considering a long-term relationship here, with commitments on both sides, I think a contract is clearly indicated. And that also goes with the next point I want to raise. *(MEL and JOHN look at one another, in a "what's-next?" vein.)*

DENISE *(continuing):* I'm thinking about loss-and-damage liability now. And my question is whether you might be interested in eliminating this from our relationship.

MEL *(cautiously):* Well, obviously the liability question is of concern to us. . . . Can I ask you what you have in mind here?

DENISE: Normally, of course, you're fully liable regardless of fault or negligence. That's standard common-carrier liability. But I'd like to consider another approach.

CHARLES *(picking it up):* Our history with these products has been that we have relatively few claims; our products are not very susceptible to damage. But we *have* had a few claims, and when we do there's usually a fair amount of money involved, because these particular products are fairly high-value. Now, we carry our own insurance, and we can extend our coverage to transportation. That way we avoid claims disputes, and of course, since this eliminates any question of your liability, it reduces your costs. But expanding our coverage adds to *our* costs, and we'd want to work out some kind of arrangement for compensation, in the form of a further rate discount.

MEL: Mm. As I understand you, you're proposing some kind of super-released-rate situation here, where we wouldn't be liable for loss and damage at all—

CHARLES *(interrupting):* Except for negligence, of course; I don't think we'd be prepared to waive our right to negligence-based claims.

DENISE: I think the law won't even allow contracting against negligence, in any event.

MEL: All right, except where we're negligent. In effect, you're talking about warehouseman's liability. And in exchange you're looking for a further discount of. . . how much?

DENISE: Five percent.

JOHN: That seems a little high, here. I mean, our experience is—well, we've handled several hundred loads for you up to now with only a couple of claims.

CHARLES: Yes, I checked that, but those claims added up to over $20,000.

MEL: Even so, I don't think we could go to 5% on this, with the low risk element. Now, if you'd be prepared to accept 2%. . .

DENISE: We can go back and forth on this, but how about simply splitting the difference at 3½%—I think that would be equitable for both of us.

JOHN: Three percent flat? *(Gets busy on his calculator again.)*

DENISE: Oh. . . all right, 3%. Carrier liability only for proven negligence.

CHARLES: Where does that put us now?

JOHN: Staying with the per-mile figures we've been discussing, I make it $1.07 a mile to New Orleans and St. Louis, $1.13 to Little Rock, $1.26 to Dallas and Oklahoma City, and $1.34 to Austin.

MEL *(firmly):* And I think that's just about as far as we can go under any circumstances; we're talking about some really exceptionally low rates here. In fact, if we're going to give you rates at this level, I'm going to ask that the contract include a secrecy clause; I really don't want to have our other shippers learning about these rates and pressuring us to give them the same.

DENISE: That's fine with us; I think a secrecy clause might be in our interest, too. In our market we don't think much of giving our competitors any more information about our costs than we have to.

CHARLES: I wonder if there isn't one more factor that might be worth discussing before we fix on these rate levels. I'm thinking about billing and credit terms. *(MEL and JOHN look at him closely, but neither says anything.)* XMZ Manufacturing's policy is to pay on time—in the case of carriers, within the standard seven-day credit period after billing. Now—

MEL *(interrupting):* We understand that, of course; and we've already taken this into account in the proposal we've offered.

CHARLES: Yes, but please hear me through. Since we're handling this on a contract basis, we both have a bit more flexibility here. What I'm suggesting is that we both might benefit from a different kind of billing arrangement.

MEL *(cautiously):* Oh?

CHARLES: We're talking about a substantial volume of traffic each week—on the order of 20 to 30 shipments or more. It's burdensome on you and it's burdensome on us to bill and pay for each shipment individually. My suggestion is that you consolidate the entire week's shipments into a single billing, to be submitted on the following Monday, say, and we undertake to pay that consolidated billing within the seven-day period—that is, by the *next* Monday.

MEL: Well, I suppose there's some merit to that idea. . . .

DENISE: And in exchange we'd ask for an additional 1% discount for payment within that time frame.

MEL: Now, there you lose me. In point of fact you'll actually be getting an extension of your present credit terms. On the shipments that move early in the week, you'll have nearly twice as much time to make payment. In the circumstances I'd think it would be more appropriate to *add* a percentage point, not deduct one.

CHARLES: Yes, but you'll be reducing your billing costs—

MEL *(jumping right in):* —and you'll be reducing your payment costs, and getting additional credit time as well.

DENISE: All right, suppose we call it a draw; you bill as we've suggested, weekly, and we pay each billing within seven days, based on the proposal we've agreed on.

MEL *(sighs):* All right. But if we're going to arrange things that way, I will have to insist on an enforcement clause concerning your payment time—a 2% penalty for any payment beyond the seven-day period.

DENISE: One percent?

MEL: One percent, then—but rigidly applicable for any late payment, even a day.

DENISE: I think we can live with that. Now, the only thing I think we have left to discuss is the timing here—how long do we want to fix the contract term?

JOHN: How about a year to start with? That seems reasonable here; it gives us some assurance of traffic for an extended period, and assures you of service for the same time.

DENISE: I think that's acceptable, although for that kind of time we'd want to include a meet-or-release clause in case—

MEL: I'm sorry, Ms. Wachower, but Victory Trucking has a very firm policy on that point. We'll be happy to enter into a contract with you, but we strongly feel that any contract must be fully binding on both sides. I'm sure you understand that we simply can't be obliged to meet the offers of any fly-by-night carrier in order to retain business for which we have a contract that's binding on our own organization.

CHARLES: How about six months, then, firm? No meet-or-release, but a short enough time that if we do get a better offer we won't have to just throw it in the garbage.

JOHN: Well. . . Mel?

MEL: We'd prefer a year, but—I think six months would be

acceptable here, in light of the heavy traffic volume that's involved.

JOHN: With an evergreen clause and a one-month cancellation.

DENISE: All right. *(Nobody talks for a moment.)*

Gentlemen, I think that about does it, then. For myself, I think we have an agreement we can both live with here, and I'm extremely satisfied. Let me buzz my secretary and we can dictate a memorandum of agreement right here, and let the lawyers put it into a formal contract later. *(She buzzes on the phone.)*

MEL: Well, I think you wound up with a bit more than we really intended to give here, but on the whole I'd agree.

Closing dialogue and credits.

8

CONTRACTING FOR CARRIERS AND SHIPPERS

A contract, according to *Webster's Third New International Dictionary,* is "an agreement between two or more persons or parties to do or not to do something. . . with mutual obligations."

This notion of "mutual obligations"—bilateralism—is the outstanding principle that underlies the entire concept of contracting. That is, a contract involves a *quid pro quo*—tit for tat, as it were—whereby one party agrees to thus-and-such and, in exchange, the other does so-and-so.*

Indeed, without bilateral commitments a contract is legally unenforceable. If one person agrees to become another's indentured servant for so many years, for example, and receives nothing in exchange, the paper is worthless in a court of law. And there must be some reasonable parity of commitments from the parties; if the indenture papers called for payment of, say, a penny a year, the courts still wouldn't accept it.

On the other hand, courts don't pass commercial judgments; one can't get out of a bad deal simply by claiming that a signed

---

*In law things are phrased a bit differently, but the effect is the same. A contract is legally deemed to be "a promise or a set of promises" by one party which is "supported by consideration" by the other. The fact that party A's contractual commitment is termed a "promise" whereas party B's is called "consideration" does not, however, alter the *quid pro quo* nature of their agreement.

contract is unfair. Following through on the foregoing illustration, if one agreed to work for another for years at the legal minimum wage, for instance, he could not later invalidate the contract by arguing (no matter how accurately) that on the open market his labors would command a far higher price.* By the same token, anyone who negotiates a disadvantageous (to him) contract can't expect the courts to bail him out.

The typical modern-day contract involves a payment of money by one party in exchange for the other's provision of products or services of (presumably) roughly equivalent value. This is the type of contract negotiated between transportation shippers and carriers.

At its simplest level, a contract need not be an extensively detailed document. The cash-register tape one receives at the supermarket is a form of contract; it says he has agreed to purchase, say, two green peppers, three tomatoes and a pound of meat for so much money. The credit-card receipt from the neighborhood filling station represents the station's agreement to sell, and the purchaser's to buy, gas at the specified price.

And in transportation the standard-form bill of lading for a single shipment qualifies fully as a legally binding contract.

Most people, however, when they talk about contracts, have something a little more in mind. As the term is generally used and understood, a contract refers to a document covering more than one transaction and/or extending over a period of time. In other words, it involves something like a hedge against the future for the contracting parties; each is, in effect, wagering that conditions ahead will (or will not) change, and that any changes will (or

*Those who are accustomed to reading and hearing about sports contracts reported in the news media may find this surprising. Because of the peculiar nature of professional sports, it is commonplace that such contracts will be "renegotiated" and otherwise treated fairly casually in terms of their binding nature. In normal business relationships, however, the situation is quite different; and in fact even sports contracts are *legally* binding on the parties, notwithstanding that in most cases professional athletes, coaches, managers, etc., are as a practical matter rarely held rigidly to their contractual commitments.

will not) take predicted forms, thereby affording him advantages he couldn't gain if he waited until later to cut a deal.

Shifting the focus again to the microcosm of transportation, a contract is deemed to be an agreement applying to more than a single shipment. Specifically, as the ICC phrases its rules, contract carriage covers "a series of shipments during a stated period of time. . . ."

Finally, a contract usually (although not necessarily from a legal standpoint) involves a written document executed (*i.e.,* signed) by the contracting parties. In transportation this is enforced by statutory or regulatory standards requiring such written documents; without the pieces of paper no contract, in law, exists between shipper and carrier.

This paperwork, therefore, constitutes the fundamental distinction between "common" and "contract" carriage, as they are known in regulatory parlance. Both, strictly speaking, involve contracts; no for-hire transportation is performed without some form of contract. But the common-carriage "contract" is merely the bill of lading covering a single shipment, whereas contract carriage involves multiple movements that may take place over a span of weeks, months, even years.

The Element of Mutuality

Getting back to the notion of bilateralism, such a contract necessarily entails commitments by both sides. Naturally, each will be striving, during the negotiations process, to minimize the degree of his own commitment and maximize the other party's. But both must—under the precepts of both contract law and transport regulation—offer at least some degree of commitment. To quote from the ICC's rules again, a transportation contract must be "a bilateral agreement, imposing obligations upon both shipper and carrier. . . ."

For the carrier's part, his commitment will normally be focused on an amalgam of price and service. His side of the contract will read roughly like a common-carrier tariff, with a "rules" section specifying the level of service to be provided and a "rates" section identifying the prices he will assess for that service. Again like a tariff (or, for that matter, an unregulated rate circular or schedule), the contract may call for varying charges depending on the level of service required for different shipments; for example, the contract may provide for "accessorial" and other special charges in addition to line-haul rates (which may themselves be predicated on any basis mutually agreeable to the parties).

The difference between a tariff and a contract (or contractual rate schedule, since rates are often set forth in separate documents attached to the contract itself), though, is that the provisions of the latter may not be unilaterally altered by the carrier during the life of the contract. In other words, the rate/service package being negotiated by the carrier today binds him for as long as the contract is in force. It need not remain constant the whole time; for example, an inflation factor may be incorporated, either based on a fixed percentage level or (more commonly) tied to some type of economic index. But these things must be provided in the basic contract; the carrier is not free, as he is with common-carrier tariffs (or circulars), to introduce changes at will at a later date.

The shipper's side of this agreement will normally deal with the quality and (especially) quantity of freight he will tender the carrier during the life of the contract. Neither need be detailed very extensively; to offer an extreme illustration, the contract may call for "freight, all kinds" with a minimum tender of but a single shipment per year "plus such other freight as the shipper, in its sole discretion, may elect to tender." Or it may be much more specific, restricted to movement of but a single commodity with high volume requirements applying both per-shipment and per-time period.

Neither is the shipper free to make unilateral changes in the contractual provisions. If, for example, he has agreed to a high volume requirement and his business suddenly falls off to the point where that requirement is no longer realistically attainable, that, to put it bluntly, is the shipper's tough luck; the contract still binds him, and he will have to pay whatever deficit-volume charges are provided.

It is only by mutual consent that a contract may be changed. Under the law, a contract is regarded as a private deal between the parties, and has (as a general matter) no further legal implications. Thus, any time both agree to renegotiate on other terms they are free to do so; no-one else (including, except in very limited circumstances, the government) has the right or authority to say them nay. Regulatory limitations that once restricted the ability of shippers and motor carriers to engage in such renegotiations are no longer are in force.

The likelihood is, however, that there will rarely be such mutual agreement. For the most part contracts, being limited to only a very few (usually but two) parties, are essentially zero-sum in nature; that is, to the extent one party is disadvantaged the other receives an equal and offsetting benefit. Just as the winning player around the poker table (poker being likewise a zero-sum game) is unlikely to return any portion of his profits to those from whom he has won them, so the "winner" in a contract negotiation probably won't agree to re-cut the deal at the request of the "loser."

Nevertheless, there are circumstances under which renegotiation may develop. Usually these will involve an excessive imbalance in the results, to the point where they are unendurable to the losing party. The shipper may find he has committed himself to such a bad deal that excessive transportation costs are drying up the market for the goods he is shipping; or the carrier may find his rates so low that they threaten him with insolvency. In such cases simple self-interest dictates some sort of "give-back" by the victorious party, since otherwise the "loser" will be obliged to abandon the field and the potential for future gains will be forever lost. And it's even possible that straightforward charity might play a role here.

Quite obviously, though it would be a serious mistake to depend on such circumstances materializing. Both parties should—must!—negotiate on the premise that the contract will remain in full force and effect throughout its stated life, and should ration their degree of commitment accordingly.

At the same time, there is one circumstance under which the provisions of a written contract are not deemed legally enforceable: where it becomes evident that the document does not accurately reflect the parties' intentions. This is identical to the situation occasionally encountered with regard to common-carrier bills of lading; if the bill of lading doesn't accurately describe the movement, the carrier is obliged to charge, and the shipper to pay, for the movement that actually took place, not as it was billed.

Legally speaking, a written contract is *prima facie* (at first glance) evidence of the agreement between the parties. That evidence is rebuttable, however, if one party can prove that the actual agreement was something else and that, either through error or purposeful misrepresentation of the other party, the contract doesn't truly embody that agreement. In such cases courts will enforce what is found to be the true agreement, rather than what was written down on paper.*

Here we find a clear distinction between common and contract carriage. Common-carrier tariffs are rigidly enforced in accordance with what's on the printed paper, regardless of what may have been intended. A rate may be ridiculously high or low because of a typographical error; nonetheless it's considered binding until changed. In a few cases the ICC has permitted deviation from published tariff rates because it found palpable errors had been made in the clerical process of publication (and usually then only when the error was in the carrier's favor); but these have been permitted only as exceptions to the rule, and then only with specific regulatory consent. More often by far, errors have been deemed uncorrectible—when it comes to common carriage.

Contract carriage, however, is not so rigorously constrained to its documentation.

Nor, at the same time, does the shipper have a legal advantage when it comes to interpreting the meaning of contract

---

*In practice, however, this will rarely happen. Legal limitations on the nature and type of evidence that may be presented to invalidate a written contract are extremely stringent; and courts are always predisposed in favor of the documented form of the agreement. Thus, neither party should depend on this process except as a last resort.

provisions. With respect to the tariffs of common carriers, provisions whose meaning is unclear or ambiguous are always, by law, construed in the shipper's favor. This follows the legal precept that any document whose contents are dictated by one party will be interpreted to that party's disadvantage in cases where the meaning is uncertain (the common carrier having, obviously, sole control of what its tariff, rate circular, etc., says).

But the law considers a contract to be a bilateral (or multilateral, if more than two parties are involved) agreement, the provisions of which may be *de jure* deemed to have been written by both (all) parties acting co-equally. It therefore aims at what it considers the most accurate interpretation without distinction as to which party is favored thereby. In other words, courts will not lean unduly toward shippers (or, for that matter, carriers) in cases where disputes about contract interpretation arise.

Transportation Contracts and the Law

Because of the formally (nominally) regulated status of most for-hire domestic freight transportation service in the United States, contracting in the industry generally must follow certain regulation-dictated standards. Failure to observe these standards renders the contract unenforceable, under the doctrine that the law may not be used to sustain agreements which themselves violate the law.

In the railroad sector this means, first, that contracts must be filed with the ICC (respecting interstate transportation) or state regulatory agency (as to intrastate movements). No rail transportation contract may go into effect earlier than the day after this filing has been made—that is, the contract copy has been physically delivered (in person, by mail, etc.) to the office designated by the regulatory body for that purpose.*

*As discussed in Chapter 3, however, this does not mean that effectiveness of the *substance* of the contract—rates, service standards, etc.—must await actual regulatory filing. Present regulatory policies permit shippers and carriers to implement negotiated contractual provisions as soon as they have been negotiated, even though the written contract itself may not even have been filed

It is important to note that this filing requirement need not compromise the secrecy of the agreement. For competitive reasons many companies, on both the shipper and the carrier side of the fence, consider it of great importance to keep their pacts secret. Indeed, in some instances one party has included contractual clauses abrogating the contract if the other reveals key information to anybody else.

The law accommodates this passion for privacy by permitting the parties, when the contract is filed, to designate just about any of its provisions as "confidential"; in such cases regulatory agencies and their employees are legally bound to honor the requests. By this means rate levels, service requirements, incentive and penalty provisions, even the identity of the contracting shipper (though not the carrier) can be held secret under most circumstances. This secrecy has been upheld in several ICC and court decisions, including a somewhat bizarre case in which a consulting firm tried unsuccessfully to force the Commission to divulge contract details under the so-called "Freedom of Information" Act.

It is necessary to add the disclaimer "under most circumstances" because there does exist the possibility that third parties might compel disclosure of railroad contract terms in certain types of legal proceedings. This could happen where one party to such a proceeding needed to know the terms in order to prepare its case and invoked a process known as "discovery." If the court or agency trying the case agreed that the information was critical, it would have to be made available.

Such circumstances, however, arise very rarely indeed; on only a handful of occasions during the first five years of the Staggers Act's effectiveness did it happen. Generally speaking, the only time a third party will be able to show a legally defensible "need to know" about contract terms is where it is challenging the

---

with the regulatory agency, signed, or even set down in writing. The only prerequisite in such circumstances is that the contract be written out, signed and filed within a "reasonable time" after the parties begin implementing its provisions.

legality of the contract itself. And the grounds on which such challenges may be mounted are extremely limited.

Any contract, of course, whether or not it is concerned with transportation, may be challenged if it is "against public policy." This is a more generalized version of the doctrine that holds unenforceable contracts which themselves entail breaking the law. Indentured-servitude contracts are held invalid on this basis, for example; so (to use a transportation-related illustration) are contracts calling for carriage of illegal drugs or other contraband—or, more mundanely, motor transportation without appropriate regulatory operating authority.

As should be obvious from the requirement that rail contracts be filed with regulatory bodies (since otherwise this requirement would be meaningless), there are also certain more particularized grounds on which railroad contracts may be challenged. These are largely holdovers from nearly a century of rail transportation being strictly common carriage (railroads were legally prohibited from providing contract service prior to 1978), and are seldom invoked; but they do warrant some attention.

No railroad may, under the law, dedicate so much in the way of resources (rolling stock, in particular) to contractual service that it can't adequately serve its common-carriage shippers as well. This is of some importance, since the law expressly permits railroads to favor their contracting shippers in allocating cars, expediting train movements, providing switching service, etc. The law empowers the regulatory agency* to reject rail contracts, or even abrogate already effective ones, if it finds this precept is being violated.

This legal provision is governed by the so-called "40% rule" adopted by the ICC (and therefore, under the Staggers Act,

---

*As of the time this was written, this regulatory power was delegated to the ICC; but there were proposals pending to "sunset" the Commission—that is, to legislate it out of existence. In such an event, however, it was evident that the rail contractual regulatory restrictions would not likewise be abolished, but rather would merely be reassigned to another agency (probably the U.S. Department of the Transportation) for administration and implementation.

binding on state regulatory bodies as well). If the carrier is dedicating more than 40% of its available rolling stock (rail freight cars) to contract service, this is to be construed as an excessive commitment. The rule applies, however, only to the extent rail cars are identified *by car designation* in a contract. If the railroad merely specifies that it will provide a contracting shipper a given quantity of empty cars for loading during a specified time period, without identifying the specific cars it proposes to use more fully, that commitment is not counted against the 40% limitation. The effect, of course, is to emasculate the law in this regard; in the first 5½ years following enactment of the Staggers Act, not a single rail contract was rejected by the any regulatory body on this ground.

Other restrictions concern themselves with the question of discrimination—but only as to certain specifically identified types of movements. Shippers of agricultural commodities and forest products (including paper and paper products) may lay claim to this protection; so may interests at seaports.

No-one else, however, is thus shielded. In any but these specified circumstances a railroad is free to give one contracting shipper an advantage over another, or even to enter into a contract with one shipper while refusing to do the same with his competitor. ICC regulations do nominally discourage such discrimination, but they are toothless; if a railroad elects to disregard them, no-one—neither the discriminated-against shipper nor the Commission itself—has legal authority to gainsay it.

(In extreme circumstances the victims of such discrimination might be able to win relief, even damages, through civil litigation on other bases, such as under antitrust law. Such a case would be extremely difficult to sustain, however, and favorable results would be by no means certain; as a general matter the law is loathe to intervene in the managerial prerogatives of businesses, including the freedom to negotiate contracts with whomever they choose under any terms they choose to negotiate.)

With these exceptions, rail contract service is totally immune from any form of government economic regulation. Rates, service, loss-and-damage liability, credit terms—*everything*—

can be negotiated to suit the parties, without the potential for regulatory restraint; even if the contract terms clearly go against what would be permissible in common carriage, regulatory agencies and the courts have no authority to interfere.

Of course, this means the parties have, essentially, no regulatory protection whatever against their own negotiating ineptitude—especially problematic to shippers who are accustomed to the umbrella of common-carriage regulation. If a shipper signs a poor deal he is stuck with it. In one early case, for example, a shipper complained to the ICC that its contract, signed but a few months earlier, called for higher rates than the railroad was now offering other shippers under its common-carrier tariff. That, said the ICC, was the shipper's misfortune; the contract was still binding.

Although the contracts of motor carriers are essentially similar to those of railroads—both are, after all, *contracts*—there are also some important differences between the two modes. The first and most obvious is that, unlike railroads, motor carriers must secure regulatory operating authority to handle most types of traffic.

In this respect contract motor carriers enjoy no different basic standing from their common-carrier brethren. That is, regulatory exemptions are the same for both forms (*i.e.*, quite limited in scope), and operating rights must be obtained from the appropriate agency—the ICC as to interstate carriage, state agencies (in most states) for intrastate—before the carrier may commence operations (*see Chapter 2).*

Once the carrier has secured his operating authority, the balance of the contracting process is similar to rail contracting, with the noteworthy exception that there are *no* special regulatory grounds on which to challenge motor carrier contracts because of discrimination. Nor, indeed, do the ICC's rules even incorporate an admonition against discrimination, as they do for rail carriers.

The privacy of interstate motor carrier contracts is assured by the fact that neither they nor the rate "schedules" (usually separate documents) of contract carriers need be filed with the

ICC. It is still required that contract motor carriers and their shippers execute written contracts, and ICC investigators may demand to see them at any time; but no-one else has access to either them or the rate schedules (if separate), and they are not kept on file at the Commission. In addition, contract motor carriers and their shippers may agree to any rate changes they choose, on, literally, a moment's notice—or even retroactively.

As already discussed, under the Motor Carrier Act of 1980 each state is free to adopt its own regulatory policies. Although some states have largely followed the Federal lead—indeed, a few have even gone beyond it to totally deregulate trucking within their borders—others continued, as of this writing, to apply restrictive controls in such areas as entry and ratemaking. Accordingly, in these states contract motor carriage may be subject to restrictions that make it a less attractive, or less readily available, option for shippers and carriers with respect to intrastate movements.

For carriers of, and shippers by, other modes, contracting is essentially similar to the above-described arrangements. No other mode, however, partakes of the railroads' regulatory limitations concerning overcommitment of resources and discrimination; and only domestic water carriers must, like motor carriers, secure operating authority from the ICC or state regulatory bodies to commence such service. Other modes* are entirely regulation-free as to contract carriage, and may negotiate any terms and conditions they choose subject only to the legal limits applicable to all forms of business contracting.

Contract Negotiation Strategies

"Why am I considering a contract?"

That's the first, and most basic, question any shipper or carrier manager should ask himself before even beginning con-

*This does not apply to freight forwarders; as discussed in Chapter 2, freight forwarders' relationships with their shippers are exclusively on a common-, not contract-carriage basis.

tract negotiations. On the answer (or answers, for there may be more reasons than one) will depend the approach he will take in the negotiation process itself.

Today's marketplace orientation of the transportation industry leaves both shippers and carriers almost totally free to select the form of their relationship—that is, whether it be that of common or contract carriage. They accordingly must analyze which of these alternatives will best serve their needs.

As a general rule, the transportation form of choice should be common carriage, in deference to the flexibility this gives both shippers and carriers in the competitive marketplace. Contract carriage should be selected only if there are affirmative reasons for using it.

From the shipper's perspective, some of the reasons for preferring contract carriage might be these:

- It may offer a significant rate advantage. Because contractual relationships generally demand an enlarged commitment from the shipper (over what may be feasible with respect to common carriage), carriers may be willing to negotiate more favorable rates if their interests are protected by the security of a contract. This may be particularly true with respect to captive rail traffic, where contractual commitments may offer carriers their only real incentive to negotiate.
- There may be a need for specialized service that is broadly unattainable from common carriers. Some types of bulk operations lend themselves to contracting on this basis; so does haulage of unusually fragile cargoes (such as the household goods carriers' "third-proviso" traffic) or other goods demanding specialized equipment or services.
- The shipper may place a high value on the security of having a "dedicated" carrier or carriers. This circumstance might obtain where there is a special premium on assured delivery schedules, for example, or where the unavailability of carrier equipment as needed would be especially problematic.
- Ancillary, non-transportation services may be wanted as part of the negotiated "package." Contract carriers—especially in the trucking sector—may, for instance, allow the shipper's

advertising material to be emblazoned on their vehicles, have truck drivers distribute promotional literature and/or take customer orders, etc.

• It may be competitively important that the shipper keep his transportation arrangements secret from competitors, customers, etc.

There may, of course, be other reasons that a shipper may prefer contract to common carriage. Whatever the reason(s), it (they) should be clearly identified and articulated within the shipper organization before the decision to seek out contract carriage is made.

For his part, a carrier may look favorably on contract carriage for such reasons as these:

• He may want a solid volume commitment from a shipper over an extended time period. Especially in these competitively pressured times, there's an obvious benefit to a carrier in "tying up" a particular shipper's traffic, to at least some extent; he need not fear the traffic will vanish tomorrow morning, leaving him with a big revenue gap to fill. Even where contracts are cancelable, typical requirements are for at least 30 days' advance notice—leaving the carrier at least some time to scramble around for replacement traffic before the contractual movements end.

• He may not want to maintain a big marketing organization, preferring to concentrate on high-volume business from a relatively small customer base. By not having to constantly "sell" to the public at large, the carrier can often substantially reduce his overhead.

• He may prefer to avoid regulatory and/or other legal requirements. Although much reduced, regulatory standards applicable to common carriers were still, as of when this was written, appreciably more burdensome than those which governed contract carriers. Merely the freedom from having to publish tariffs, comply with credit and other regulations, etc., may be of value to a carrier. Additionally, contract carriers are not subject to the law governing common-carrier liability for loss and damage (which applies even to otherwise-unregulated airlines).

• He may wish to keep confidential especially advantageous rate/service offerings he establishes for particular shippers. If he serves those shippers as a common carrier his rates will be exposed to public view in the form of tariffs, rate circulars, etc.; thus, he may face the choice of either extending the same deal to other shippers or offending them by refusing to do so. With a contract, however, those other shippers will not be so readily able to learn the terms under which the carrier is providing the service in question.

• He may elect to provide contract carriage simply to satisfy the predilection of a particular shipper. For the most part it will be the shipper, as purchaser, who makes the choice of form; and some carriers simply "go along" to accommodate their larger customers' wishes in this regard.

Or, again, he may have other reasons, or combinations of reasons, of his own.

Each of these reasons, each different combination of reasons, will call for a different negotiating approach. The shipper who principally wants specialized or dedicated service will obviously take a different tack during negotiations than will the one whose objective is mainly to secure reduced rate levels. Carriers seeking to build up overall traffic volume won't approach things the same way as others who may be trying to fill empty traffic lanes. And so forth.

The shipper or carrier manager is thus obliged to define, as clearly and *specifically* as possible, the reason(s) why he, in his and his company's particular circumstances, is looking to establish a contractual relationship. This will give him the necessary basis for all that must follow.

Once this definition has been achieved, the preparation for contract negotiations is largely identical to that for common-carriage negotiations (*see Chapters 4 and 5).* But in the context of contract carriage it is especially important as a preface to contract negotiations that the parties research one another *thoroughly.* Any shipper who agrees, in contractual form, to tie himself to a carrier about whom he knows little or nothing. . . or any carrier who does the same with an unfamiliar shipper. . .

richly deserves any problems he may have. Where common carriage is involved, failure to research one's opposite number can be excused on the ground that, if the relationship proves unsatisfactory, one is at least not deeply committed to it; but since contract carriage does involve at least some degree of commitment, it is nowhere near so easy to escape from even a thoroughly bad relationship once the papers have been signed.

Key Features of Transportation Contracts

As with common carriage, the single most important factor in any transportation contract negotiation will probably be the specifications of the traffic (*see Chapter 5).* To recapitulate briefly, these factors include:

- *Commodity considerations:* commodity type and/or classification, form of packaging, handling characteristics, "stowability", density, hazardous nature of the commodity(ies) (if any), special in-transit security requirements (if any), value, fragility, susceptibility to theft and pilferage, susceptibility to contamination and infestation, etc.
- *Geographic considerations:* origin(s) and destination(s), length of haul, territory(ies) or region(s) involved, etc.
- *Volume considerations:* individual shipment volume, aggregate volume, periodicity or seasonality of movement, etc.

Every transportation contract will necessarily contain specific provisions relating to these three fundamental factors. That's not to say each and every distinction must give rise to a separate clause; averaging techniques are often employed to develop, for example, "Freight All Kinds" (F.A.K.) rates (based on an average of the commodities the contracting shipper wants moved), ZIP-Code-based rate structures (reflecting average distances between points in three-digit ZIP Code areas), and the like. But these are merely distinctive treatments of the more particularized factors, which form the bases for development of those averages.

In contracting, special attention must be given to the subject of aggregate volume. During the days of stricter transport regulation, contractual relationships in transportation (principally motor carriage, with a modicum of domestic water and some air service) had a different character than in most other economic sectors. That is, contract carriage was in many cases merely an expedient to avoid the tougher regulatory standards applying to entry, ratemaking, etc., for common carriers; and in these circumstances relatively little attention was paid to shipper commitments of substantial traffic volume. The typical motor carrier contract, for example, would oblige the shipper to tender only a very minimal amount of traffic to the carrier; he had the *option* of shipping high volumes of traffic via that carrier, but was not contractually required to do so.*

In the post-1980 deregulated environment, however, the situation has changed considerably; carriers can today enter the marketplace with equal facility as either common or contract carriers, and rate regulatory burdens have been substantially lessened (or in some instances eliminated altogether). Accordingly, carriers are increasingly insisting that shippers commit to substantial volumes of traffic before they will consider a contractual relationship.

The effect of this is to bring the two parties' commitments more into parity, as is generally the case in other economic sectors. In all likelihood this trend will prove, in marketplace terms, irreversible; that is, carriers will grow increasingly unwilling to commit themselves contractually unless they receive an equivalent commitment from the shipper. Accordingly, volume commitments by shippers seem certain to play an ever-more-important role in transportation contracting as time goes along and deregulation takes a firmer hold in the market.

Next in importance in transportation contracting is the quality and particulars of the service to be provided. Transportation service, in the modern-day world, entails much more than simply the hauling of a load of goods from this point to that. Among the other factors that need to be considered are:

---

*Such agreements are known as "minimum-volume" contracts.

• *Pickup service.* Normally the carrier will pick the goods up at the shipper's facility (or other point designated by him). In what fashion, with what frequency, on what advance notice, etc., is this to be done? If rail service is involved, the contract will usually specify switching schedules; in the case of motor carrier service it involves decisions as to whether power equipment is to await the loading and release of trailers or whether they are to be "dropped" (left without the power unit) on the shipper's premises for loading at the shipper's convenience. Sometimes pickups will be strictly on a scheduled basis, and the schedule should be established in the contract; in other cases the shipper will notify the carrier when a shipment is planned, and notification requirements (especially advance-notice standards) need be spelled out.

• *Delivery.* Are goods to be delivered to the consignee's premises at the carrier's convenience or at the consignee's? Is the carrier obliged to telephone in advance for a delivery "appointment" (commonplace in motor carriage)? Will rates vary depending on whether the consignee has an outbound shipment he can load aboard the same vehicle once the incoming goods have been unloaded?

• *Free time and demurrage/detention.* How long will the carrier be expected to wait (or leave his equipment) at the pickup and delivery points before extra demurrage or detention charges are assessed? Will "averaging" of demurrage or detention be permitted, so that shippers or receivers may in effect "save up" free loading/unloading time from equipment released early? Ensuring that the contract covers these points protects the carrier against having his assets tied up for days, even weeks at a time because the shipper or consignee is dilatory about releasing it. At the same time, it may be worthwhile anticipating exceptional circumstances that might warrant special treatment here, such as labor strikes or inclement weather that prevent prompt loading or unloading due to no fault of the shipper or consignee.

• *Loading/unloading.* This has pertinence mostly to motor carriage, since in most (although not all) instances other modes do not offer loading or unloading service at shipper/consignee premises. Who is to do the loading at origin, who the unloading at destination? If the carrier has responsibility for either activity, the contract should also roughly define what constitutes loading and what unloading. Are carrier employees expected to walk deep into the shipper's facility to find the goods and then manhandle them back to the vehicle, or will they be staged nearby on the loading dock? Does unloading constitute merely stacking the goods near the vehicle's tailgate, or is more contemplated—"inside" delivery, sorting and segregation of the freight, etc.? In the case of bulk shipments that require special hoses and connectors for loading and unloading, who is to provide that equipment?

• *Transit time.* What standards are to apply, and with what guarantees (backed by what incentives and/or penalties)? Not all contracts include transit-time specifications, but failure to do so not only leaves the shipper poorly protected from carrier service failures but also encourages shipper-carrier disputes which may prove difficult to resolve. Where service between a limited number of point pairs (origins/destinations) is involved, it may be desirable to specify transit times on a point-to-point basis; in other cases (this approach is frequently used in motor carrier contracts) transit times may be specified on a zone or mileage basis.* For obvious reasons the contract should generally incorporate some form of incentives or penalties to encourage adherence to the transit-time standard(s) specified; most common are monetary incentives/penalties, but an alternative approach is to provide for automatic cancellation or abrogation of the contract at the shipper's sole option if the standard(s) is (are) not met on a fixed (high) percentage of shipments.

*Shippers are advised to exercise care, however, that any contractual specifications for motor carrier transit times can practicably be met by carriers within the scope of existing hours-of-service and other safety rules. First, transit-time requirements that can't be met within these standards would be deemed legally unenforceable on the ground that they were contrary to public policy. Second, a shipper who insisted on transit times beyond what the carrier could accomplish within the confines of safety regulations might be deemed collaterally liable for damages resulting from accidents in which the carrier was deemed at fault.

• *Co-loading.* This has application mainly as to motor carrier contracts calling for movement of less-than-truckload shipments. In such cases it will usually be a good idea for the contract to spell out what restraints, if any, are to be imposed on the carrier's co-loading of other shippers' goods aboard the same vehicle at the same time, including, particularly, whether vehicles are or are not expected to move under seal.

• *Protective service.* Where goods are to be protected against temperature extremes (or, in the case of frozen or chilled products, significant changes in temperature), the contract should specify this with some particularity. Generally it is a good idea to identify the precise maximum and/or minimum temperature, allowable variation in temperature range, etc. Other types of required protective service, such as protection against moisture or humidity, should likewise be spelled out.

• *Accessorial services.* What other services (if any) is the carrier to provide in conjunction with his transportation of the goods? Multiple stop-off pickup/delivery? Sorting and segregation of the freight? In-transit storage of shipments? Returning pallets or other used shipping containers or devices? Collection, order-taking and the like from customers? Securing governmental (or other) inspections? Obtaining required special governmental permits? The contract should identify everything the carrier will be doing with regard to the shipments (or may do with regard to some of them) and what, if any, additional charges are to be assessed for each such service.

• *Shipment documentation.* When the shipper turns goods over to the carrier, he will want some form of receipt for them; similarly, the carrier will need some sort of shipping instructions. What format is to be used for these purposes? It is in many cases convenient to rely on the standard-form bill of lading (although, if this is done, care should be taken to strike out any of the "fine-print" clauses on the back of the pre-printed document that don't apply), but the contracting parties may specify otherwise if they choose. The same holds true for delivery receipts, loss-and-damage claims and the like; all documentation should be standardized by contractual provision.

• *Contingency provisions.* What happens if all does not go smoothly in the course of a particular shipment's transportation? Suppose the consignee unexpectedly refuses the goods, leaving the carrier stuck with them in his possession? What if the carrier forgets to collect a C.O.D. amount? And so forth. It's the unexpected that gives rise to most contract disputes; the intelligent shipper or carrier will seek to anticipate such problems and, to the extent possible, make provision for them in the contract.

• *Loss and damage liability.* This is a particularized type of contingency provision that should *always* be incorporated in every transportation contract. The standard law of freight loss and damage applies only to common, not contract, carriage, and so cannot be relied on to handle such situations (unless, of course, the contract provides for its application). The contracting parties are thus entirely free to specify the conditions under which the carrier will be liable for loss and/or damage, limitations on the amount of his liability, etc., pretty much at their mutual option. The only condition that isn't acceptable is where the contract seeks to exonerate the carrier from liability even where the loss or damage was due to his negligence; such contractual provisions are legally unenforceable. (For that matter, the contract likewise may not legally enforce on the carrier liability for the consequences of the *shipper's* negligence, as in packing, loading, etc.; such "contracting against negligence" is impermissible no matter who benefits.)

• *Billing and payment.* Although greatly relaxed at the beginning of 1985, ICC regulations still restrict to some degree the extension of credit by surface common carriers, dictate the form and timing of carrier billing, etc. Once again, however, these rules have no application to contract carriage. Thus, it's up to the parties themselves (if they choose) to provide for such things. Carriers may offer prompt-payment discounts or assess service charges for late payment; billing may be on a periodic basis for an aggregation of all shipments made during the period; billing may even be equalized over an annual period to allow for high seasonal variances in shipment volume; and so forth.

• *Dispute settlement.* This is one of the most commonly

neglected areas of contracting, yet is (or should be) one of the most important. The contracting parties may provide, if they like, for arbitration (binding or non-binding) or other means of resolving disputes. It is even possible to treat different types of disputes differently, providing, for example, for one party to arbitrate loss-and-damage matters, another to handle rate/service problems, etc. The obvious advantage of a dispute-settlement provision of this type is that it avoids the cost, red tape and (above all) the delays incumbent in court litigation, which, in the absence of such a provision, will be the only available means for handling disputes.

• *Duration.* When does the contract commence? When does it terminate? What provisions (if any) are to be made for its continuation beyond the nominal termination date? What about cancellation (with or without penalty) if either party becomes unhappy with it during its term? The contract should of course cover all these points fully; all are completely at the discretion of the contracting parties except that (*as noted above)* railroad contracts may not take formal effect until one day after they have been filed with the ICC and/or state regulatory agency.*

• *Secrecy.* If confidentiality of the contractual terms is important to either party, a provision to ensure it should be incorporated in the body of the contract itself. The only limitation applies, again, to rail contracts, whose terms must be revealed to regulatory bodies (though the carrier may, when he files the contract, identify the particular provisions he does not desire be revealed to anyone else). Provisions calling for secrecy are admittedly difficult, if not impossible, to enforce, since it will rarely be possible to identify the source of a "leak" if information does get out; but they nonetheless may discourage idle talk that might

---

*In this context mention should also be made to the so-called "evergreen clause" found in many contracts—a provision to the effect that, once the initially specified time period has expired, the contract will be self-renewing in perpetuity unless canceled in writing by one of the parties. This allows the parties to continue their relationship beyond the time period originally contemplated provided both remain satisfied with it, without obliging them to go through the effort and paperwork of formally renewing the contract.

inadvertently disclose competitively sensitive information.

• *Price.* Finally, of course, the contract must specify the rates and charges to be assessed. Most carriers elect to do so in a form fairly similar to that employed in common-carrier tariffs, since both they and their shippers are familiar with this format. However, any approach is acceptable provided only that it is clear and unambiguous. If the contract is for an extended period of time (more than, say, a few months), this part of the contract should also include appropriate provisions for rate indexing to allow for inflation.

In addition, some shippers may want other types of protective provisions to ensure that the contract—especially if it covers an extended time period—will not hamstring them inappropriately. Two such provisions sometimes employed are:

• *The "most favored nation" clause.* This provides, in effect, that the carrier may not offer the contracting shipper's competitors more favorable deals. Such a clause guarantees the shipper that the contract will be amended, if necessary, to afford him no worse treatment than his competitors receive, should they at some subsequent time negotiate their own contracts with the same carrier (or employ that carrier's services under a common-carrier tariff).

• *The "meet or release" clause.* By this the carrier agrees that, should other carriers in the future offer the shipper a better deal (by means of common-carrier tariffs, contractual agreements or both), he will either (a) meet that deal in all its particulars, or (b) release the shipper from any remaining contractual commitments so it can take advantage of the competing offer. Such provisions commonly impose on the shipper the burden of documenting the *bona fide* nature of the competing offer, in some instances even specifying the particular documentation required.

All of this would obviously make a formidable document if it were spelled out fully. In most instances, however, that is not done; rather, the contract itself covers only a few key points, and deals with the others through reference to other documents—existing carrier tariffs (especially mileage guides, commodity classifications and the like), standard-form bills of lading, separate rate schedules, even the various statutes and regulations applicable to common carriage. The contract proper, thus, may run as little as a few pages length, specifying only those points not adequately covered in other, referenced documentation.

What is important is not the *manner* in which the contract deals with its provisions—whether within its own "four corners," or by ancillary documents such as rate schedules, or by reference to other existing publications—but that it does deal with them *somehow.* The two biggest pitfalls of contracting are, first, not making sure the written contract fully and accurately reflects the scope of the contracting parties' agreement, and second (and even more basic), failing to reach agreement at all on key points affecting the real-world activity the contract is meant to cover.

It must always be remembered that a contract represents an accommodation between two (or more) parties whose interests tend to be at least somewhat opposed. To put it very basically, the shipper wants to get a high level of service for a low price and the carrier wants to get a high level of revenues for a low operational output—objectives that clearly tug in opposite directions. Reconciling them is certainly not impossible (if it were, shippers and carriers would be unable to do business with one another), but it requires time, skill and, above all, a freedom from external pressures while the process of negotiating toward that accommodation is going on.

This last element, of course, is precisely what's missing in the event that unprovided-for contingencies arise during the operational life of the contract. Each side naturally takes an initial view of things solely from his own perspective; but the ongoing existence of both the contract itself and operations under it creates intense pressures to impede any even-handed resolution of the problem area.

This is why renegotiation will rarely (if ever) prove a satisfactory means of dealing with contractual uncertainties and/or omissions. The shipper who fails to specify a particular service he needs can expect to wind up paying through the nose for it after a renegotiation; the carrier who neglects to detail billing and payment standards shouldn't be surprised if his collections are forever running 30 days (or more) behindhand. As a last resort, renegotiation during the life of a contract will usually be better than silently enduring the problem; but the real answer is to plan ahead intelligently at the time the contract is *first* being negotiated and cover all reasonably foreseeable contingencies *then,* before day-to-day pressures have distorted the bargaining posture.

The Ethics of Contracting

There are two major reasons to be concerned about ethics in contract negotiations. The first, of course, is straightforward personal (and/or corporate) integrity. Most individuals and companies make it a point to do business in an honest, above-board fashion simply because that is the way in which they wish to do business.

Secondly, virtually all individuals and firms—even, ironically, those who don't themselves hold to very high ethical standards—expect *others* to behave toward them with honesty and integrity, and try to avoid dealing with those who don't. Since a contractual relationship is, in a very real sense, a form of business partnership, it's easy to see how unethical practices can sabotage the relationship.

It should be borne in mind that the contract has yet to be written that is 100% proof against one of the parties undermining it; if either really wants to do so he can make things thoroughly unpleasant (at the least) for the other. The only truly successful contract is one that both parties *want* to keep—and this doesn't leave room for unethical practices by either that engender distrust and dissatisfaction in the other.

In addition, of course, managers who have had unhappy experiences dealing with a particular individual or firm are prone to be a bit gossipy about it. Thus, those who act unethically will soon find themselves developing an unsavory reputation, which is likely to impair their ability to develop future business relationships with others, as well.

Ethics in business contracting—whether in transportation or any other economic sector—involves, principally, honesty. That doesn't mean either party must open up all his secrets during contract negotiations; that's not honest, merely foolhardy. It *does,* however, mean that when negotiators do say something they speak the truth as they see it, and avoid deliberate falsehoods or "clever" half-truths designed to convey a false impression.

Thus, it's ethical for a carrier not to reveal his costs of handling a particular movement, but unethical for him to purposely overstate costs. It's ethical for a shipper to avoid discussing the total volume of his traffic, but unethical for him to say or imply that he will be moving more traffic than he actually plans. And so forth.

Most important, it is *highly* unethical for either party to take advantage of the other's mistakes to incorporate into the written contract wording that carries a "hidden" meaning—a meaning, that is, of which the other side is not aware. The purpose of a written contract, it must be remembered, is merely to record in indelible form the "meeting of the minds" that has been reached between the contracting parties. It is obvious, however, that if one party doesn't really comprehend the full meaning of particular portions of the contract, the document does *not* in fact accurately reflect any true "meeting of the minds."

Suppose, as an illustration of this sort of thing, that the carrier and shipper agree to include in the contract a clause exonerating the carrier from liability for damage due to defective packaging. During the course of negotiations the shipper has explained to the carrier precisely how his goods will be packaged, and has even shown a sample. The carrier, although knowing (by dint of his expertise) that this particular packaging isn't sturdy enough to protect the goods from substantial risk of in-transit damage, says nothing—nor does he point out that, if he were to haul those goods as a *common* carrier, his knowledge of the packaging deficiencies would preclude him (under the common law) from claiming this defense against liability. This would be classically unethical behavior on the carrier's part. (Similar illustrations could, of course, also be offered from the other side of the bargaining table.)

Necessarily, ethical behavior involves drawing some fairly fine lines. Especially in negotiating contracts, however, it is important to draw those lines very carefully indeed, so that behavior not only *is* ethical but is *seen,* by the other party, to be ethical. A contractual relationship is, by definition, intended to last for an extended period of time (at least the life of the contract); and one that starts out in distrust on either side can scarcely be expected to flourish.

The Negotiating Process

Contract negotiations are among the most difficult to conduct, for the simple reason that *everything* is negotiable. There are only the most minimal restraints on what two consenting parties may negotiate privately with one another, so that the shipper must consider, and deal with, virtually every aspect of the transportation rate/service package during the negotiations process.

Negotiations may be handled by either of two different means. First, of course, they may be accomplished through one-on-one bargaining, just as with common carriage. In such cases the negotiations process is essentially the same as where common carriage is involved (*see Chapter 7),* the only significant difference being that there's a great deal more that must be negotiated.

Secondly, a competitive-bidding process may be employed. That is, the shipper requests proposals from various carriers for a particular quantum of service, the carriers submit their bids on a "blind" basis, and the contract is awarded to the low bidder (*see Chapter 9).* Again, of course, a wealth of detail must be incorporated in the shipper's "request for proposal," and considered by the carrier in making his bid, because of the high degree of negotiability of contract terms.

Whichever method is employed, at the conclusion of the negotiations process the agreement must be put into final contractual form. In general, the parties should agree on which of them is to do this. The choice is largely immaterial, since both will have to review and sign it; in some instances the carrier will draft the contract, in others the shipper will.

Quite obviously, whichever party does the job should either have the document prepared by an attorney or, at the least, use a standard-form model developed by one. Since the law leaves interpretation of contracts to the court system, it's clearly important that they incorporate proper legal terminology and structure, which is extremely exacting in its use of words and phrases.

It's all very well to say (as officials in the Administration of President Carter so often put it) that legal documents should be written in "plain English"; and certainly there is room to eliminate and/or clarify some of the more archaic and obfuscatory terminology of the law's traditions. But the fact is that "plain English," as used in ordinary conversations, correspondence, etc., simply isn't precise enough to convey the unambiguous meanings required for legal matters.

Indeed, almost any variance in the way words are used, or even punctuation, tends to make lawyers extremely nervous. The number of cases in the annals of contract law that have hinged on the insertion or omission of a conjunction, the tense of a verb, even a misplaced comma, is formidable. It is obviously important to use terminology whose meaning is certain (having, over a period of many centuries, been interpreted in a consistent manner by the courts) in order to keep the contract free from ambiguities that could lead to needless disputes; and lawyers are the ones who know that terminology.

However, lawyers can only see to the legal technicalities of the contract; its contents are the province of operating personnel of the two parties. Any draft contract should thus receive a full review by *all* affected operating departments. For the shipper this means Traffic, Marketing in the case of outbound traffic, Purchasing and/or Production for inbound traffic, Accounting, and perhaps others. For the carrier review should be by both Sales/Marketing and Traffic and, depending on the carrier's internal structure, perhaps such other departments as Commerce, Rates, etc.

Obviously each party should conduct such a review irrespective of which of them prepares the document itself. In addition, the individuals who actually took part in the negotiations (if not already covered in the above discussion) should also be afforded an opportunity to look over the documentation, to make certain it comports with what was agreed.

The only remaining step is merely for each party to sign the contract. It is then (if necessary) filed with the appropriate regulatory agency(ies), and goes into effect.

Contract Administration

The contractual process, however, does not stop with the signing of the contract—although far too many people think it does. There remains the critical function of contract administration.

In *Don Quixote,* Cervantes warns that "a wise man will. . . not venture all his eggs in one basket." More recently, however, Mark Twain advised, "Put all your eggs in the one basket, and—*watch that basket."* Contract carriage necessarily entails (for both carrier and shipper) putting, if not all, at least quite a few of one's eggs in a single basket; and contract administration represents the necessary basket-watching.

The art of contract administration lies in identifying contractual problems—actual or potential—as early as possible, and then taking steps to prevent them from getting serious. To accomplish this, the contract administrator must constantly be asking himself three questions: (1) Am I holding to my part of the bargain? (2) Is the other side holding to *his* part of the bargain? and (3) Are there things which need to be dealt with that weren't fully covered in the original bargain?

Most important—and yet, ironically, most often neglected even by those who conscientiously try to practice contract administration in other respects—is the first of these. Many firms monitor the performance of the other party, but relatively few look seriously at their own actions; they take the view that "it's up to the other guy" to protect his own interests.

Such an attitude is extremely short-sighted. To begin with,

not every carrier or every shipper will dedicate much time or effort to contract administration, meaning that often problems won't be uncovered until after they've already become serious. Second, even among companies that do have sound contract administration approaches, there will be some that are reluctant to register complaints until after matters have got badly out of hand. And third, the notion that either party cares not at all for its own contractual commitments is clearly detrimental to the kind of business partnership that is embodied in a contractual relationship.

Having committed himself, in writing, to doing particular things in particular ways, anyone signing a contract should make every effort to honor those commitments. Review procedures should be established to ensure that this is being done; and if a party (again, shipper *or* carrier) finds he is falling short of his obligations, he should immediately take such remedial steps as may be necessary.

Scrutinizing one's own behavior in connection with the contract also lends added standing to one's examination of the actions of the other party. If one is not keeping his part of the bargain but complains to another about what *he* is doing, the second party is naturally going to pay a lot less heed to the complaints than if the first *were* doing what he was supposed to be doing. Indeed, as in every kind of human relationship, it is all too easy for contractual disputes to deteriorate into a kindergarten-level sort of "But he hit me first!", as each party excuses his own misfalcations by indignantly citing the (unrelated) actions of the other.

There are two phases of monitoring the actions of the other party to the contract. The first is *review* of what the other party is doing; the second is *communication* of any problems to that other party. The second, though often slighted in discussions of contract administration, is at least as important as the first.

The review process should be a rigorous one with an attentive eye to the precise letter of the contract. If, for example, the contract provides for delivery before 9:00 a.m. and the carrier actually arrives at 9:05, the shipper should take note; if allowed free time is two hours and the vehicle is actually released five minutes late, the carrier should do likewise. These are obviously petty matters, certainly not violations of at least the spirit of the agreement; but petty things, untended, have an inexorable way of growing into big ones, and the time to deal with such problems is before this has a chance to happen.*

Communication of such problems—or, for that matter, much larger and more important ones—to the other party must be handled artfully. The manager who charges in like the proverbial bull in a china shop, rattling the walls with invective and threats, is more likely to alienate his contractual partner than to achieve effective results. What's wanted are tact, courtesy and, above all, a willingness to let the other side "save face" whenever possible.

This emphatically does *not* mean overlooking even seemingly trivial violations of the contract terms. That five-minute-late delivery or equipment release should be pointed out to the other party. However, the tone and manner employed may be casual and friendly; the effort here is not so much to cast aspersions for past failures, but merely to let the other party know that what he's doing is being watched.

Even as to bigger problems, fault-finding should be avoided at all costs. Communications that take the tone "we have a

---

*And if and when such problems *do* grow—the 9:05 delivery becomes a 9:30 one, then 10:00, then 11:00, etc.—failure to register complaints about the early failures makes it increasingly more difficult to complain successfully when matters get worse.

To illustrate from history, former President Eisenhower missed a golden opportunity along these lines when the American U-2 spyplane piloted by the late Francis Gary Powers was shot down over Russia in the late 1950's. For several years these high-altitude craft had been overflying the Soviet Union on intelligence missions, unremarked publicly by Soviet authorities who did not want to admit to the world that their aviation technology was so backward that they could do nothing about the situation. Finally, when the Russians at last developed weapons systems that were capable of attacking the U-2's and actually shot one down, America embarrassed itself by initially denying that the destroyed plane was its—an embarrassment compounded when the Russians actually produced pilot Powers, who had parachuted out of his doomed plane, as proof positive of their claims.

A good contract administrator might have advised President Eisenhower differently. "*Of course* we've been overflying your territory," the President might have said blandly. "We've been doing it for years. You never complained about it, so we thought you didn't mind!"

problem" are generally going to be a lot more effective than those asserting (or implying) that "*you* have a problem." The goal, after all, is to fix the problem, not exact psychological restitution.

Along these same lines, face-saving should be permitted, even encouraged, where possible. If the carrier lamely excuses his two-day-late delivery by blaming a recalcitrant driver who has allegedly since been fired, the shipper's traffic manager should express understanding and sympathy—waiting until *after* he hangs up the phone or part company with the carrier's officials to make the obvious remarks about poor management practices. And so forth, as to both shippers and carriers. In particular it's good tactics to let the individual to whom the complaint is addressed clear himself of personal blame, even if the excuses are palpably fictitious.

Civility, however, is not synonymous with being wishy-washy. If a problem has occurred for which the contract provides some penalty, in most circumstances the aggrieved party should invoke the penalty. This should certainly be done politely, but it should also be done firmly. If the penalty is to be waived, that should be done only in exchange for appropriate concessions from the offender—and a mere "Sorry about that, won't let it happen again" is scarcely enough (since, under the contract, it wasn't supposed to happen even once).

Nor should face-saving be allowed to go to extremes. If the problem is a recurrent one, or the excuses offered leave the door wide open to recurrence, they must be brushed aside. *One* two-day-late delivery can be blamed on a lazy driver; half a dozen clearly reflect managerial problems on the carrier's part, which the carrier must be pushed into dealing with. The same, once more, holds true on both sides of the fence. Where letting the other party save face conflicts with the basic contract-administration goal of solving or preventing problems, the latter, of course, must take precedence.

Quite apparently, good contract administration demands a fair ration of diplomacy on the part of the individual(s) to whom this task is assigned. He (they) must judge when, where and, above all, *how* to exert the necessary pressure on the other party to ensure contract compliance—factors that will vary substantially from one situation to another.

Even the most skillful contract administration will not be enough to solve all problems, though. The sad fact is that some companies and individuals sign contracts to which they have absolutely no intention of adhering, or make virtual careers out of finding obscure loopholes in documents they have signed, or simply allow what is hypocritically known as "situational ethics" to persuade them to violate contract terms they find uncomfortable. No matter how diplomatically he tackles the problem, the other party finds himself up against a brick wall in attempting to resolve the problem area. And genuine differences of opinion can also crop up.

Where reasonable efforts to handle the problem have failed, the obvious recourse is litigation or, if the contract provides for it—or, even without such a contract provision, the other side agrees—arbitration. Most of the time this will be the appropriate "remedy of last resort" for irreconcilable conflicts.

On rare occasions, litigation or arbitration simply won't be practical. Because of backed-up court calendars and the possibility of multiple appeals, it can take years to complete a lawsuit in equity; even arbitration, much quicker, can drag on for months. Meantime, it may be that a critical real-world problem is continuing, exacting a serious toll on one of the contracting parties.

In such circumstances the courts recognize the validity of what is loosely termed "self-help"—unilateral actions taken by one party to offset damages he's incurring as a result of what the other is doing. For example, the party being injured may simply refuse to continue abiding by the contract, leaving the other to pursue legal remedies if he chooses. Or the shipper may withhold payment of disputed sums, or the carrier may decline to deliver shipments until he has received what he deems to be proper payment for them, or. . . .

Self-help is a very tricky proposition, legally speaking, for the party seeking to exercise it. Great care must be taken to use it *only* (1) where the delays of legal remedies will be crippling, and (2) in appropriate measure based on the injury allegedly done. In other words, if the law can solve the problem, litigation is the answer; if legal delays will aggravate the problem to unacceptable dimensions before it can be solved, one should take the least action against the other party consistent with mitigating or repairing the damage that has been done.

And, above all, one should *always* first seek legal advice before attempting any self-help. Breach-of-contract lawsuits can result in serious financial awards to the aggrieved party; and, in the law, two wrongs not only don't make a right, they leave *both* parties potentially vulnerable to such lawsuits.

The third and final aspect of contract administration is anticipating potential problems not covered by the contractual document before they become critical. If something was left out of the contract, the time to deal with it is as soon as possible, while opportunities for renegotiation have not yet been compromised by the pressures of day-to-day operations—*not* after things have already gone so badly awry that one of the parties has been put completely in the driver's seat, able to demand just about anything he wants in exchange.

In other words, the contract administrator, in this role, is second-guessing those who negotiated the agreement. He should always be alert to the possibility that something was inadvertently left out originally, and will have to be corrected with further negotiations even while the contract is in force.

To neglect any of these three aspects of contract administration is to risk serious problems. Neither shipper nor carrier can afford the risk that the contract will not work as intended, and that his goals in negotiating the document in the first place will therefore not be met.

It should always be borne in mind that a contract is a two-edged sword, offering both benefits and drawbacks. The goal of the entire negotiations process, beginning with the initial decision and running all the way through contract administration and the life of the contract itself, should be to maximize the benefits and minimize the drawbacks. Only thus can the "win/win" result so critical to a successful business relationship, in which both parties perceive themselves as receiving something of value in exchange for what they are contributing, be attained.

Appendix
CONTRACT FOR CARRIAGE

This is an agreement and contract between [name of carrier] (hereinafter referenced as "Carrier"), and [name of shipper] (hereinafter referenced as "Shipper") for Carrier to provide certain transportation services for Shipper as hereinafter specified.

(1) STATUS OF CARRIER: Carrier represents that it is an interstate for-hire motor contract carrier operating under authority granted by the Interstate Commerce Commission to provide transportation service on behalf of Shipper in interstate or foreign commerce, and fit, willing and able to provide service as contemplated by this contract. Carrier further represents that it is in compliance, and shall during the term of this contract remain in compliance, with all applicable laws and ordinances of all governments (Federal, state, local and, to the extent the operations governed by this contract involve operation in foreign commerce, foreign) having jurisdiction over its operations, and with all applicable regulations of the Interstate Commerce Commission, the U.S. Department of Transportation, and any other governmental agency(ies) having appropriate jurisdiction.

(2) COMMITMENT OF CARRIER: Carrier agrees to provide such transportation service as Shipper may desire pursuant to the terms and conditions set forth herein. Carrier further agrees to maintain, operate and keep a sufficient and adequate amount of transportation equipment to provide prompt and efficient transportation service for Shipper upon reasonable request therefor by Shipper, as defined hereinbelow. This transportation service shall be performed solely by employees and/or agents of Carrier, and Shipper shall have no control or responsibility with respect to such persons. Carrier shall perform its obligations under this contract as an independent contractor, and not as an agent or employee of Shipper; and hiring, terms of employment, and discharge of employees/agents performing transportation service pursuant to the provisions of this contract shall be in the sole and exclusive control, and shall be the responsibility of, Carrier.

(3) COMMITMENT OF SHIPPER: Shipper agrees to tender Carrier a minimum of one million (1,000,000) pounds of freight, in truckload increments, for transportation during the initial period of this contract; and a minimum of 100,000 pounds of freight, in truckload

increments, for transportation during each month in which this contract is in force and effect subsequent to the initial period of this contract. Shipper further agrees that, should it fail to tender Carrier the minimum volume of freight specified hereinabove during either (a) the initial period of this contract, or (b) any subsequent month in which this contract remains in force and effect, it shall be liable to Carrier for liquidated damages in the amount of five dollars ($5.00) for each 100 pounds, or fraction thereof, of shortfall below the above-specified minima.

(4) REASONABLE NOTICE: The parties agree that the term "reasonable notice," as used herein, shall be construed to mean that Shipper must notify Carrier of its intent to make shipment, and the number of vehicles needed to accommodate the volume of freight being shipped, at least one (1) day, exclusive of Saturdays, Sundays and holidays, prior to the date on which such shipment is contemplated. Such notice must be received by Carrier, by telephone, telegraph or in writing, prior to 5:00 p.m. on the day on which it is given; notice received after that time shall be deemed as having been given on the next day, exclusive of Saturdays, Sundays and holidays. In the event Shipper fails to provide such notice as specified by this paragraph, Carrier agrees to make its best efforts to provide equipment for the transportation of the freight in question, but shall not be deemed to be obligated to do so under the terms and conditions hereof.

(5) TRANSIT TIME: Carrier agrees to provide transportation service under the provisions hereof according to the following schedule:

DISTANCE BETWEEN ORIGIN AND DESTINATION (miles)	TRANSIT TIME (days)
450 or less	1
Over 450 but less than 901	2
Over 900 but less than 1,351	3
Over 1,350 but less than 1,801	4
Over 1,800 but less than 2,251	5
2,250 or more	6

All mileages shall be measured according to actual distance over the route of movement. For purposes of this paragraph, where loading of Carrier's vehicle is completed, and the vehicle is released by Shipper, at or before 12:00 noon, it shall be available for delivery at destination at any time during normal business hours on the day identified in the foregoing schedule of transit times; where loading is completed, and/or the vehicle is not released by Shipper, after 12:00 noon, it shall be available for delivery at destination on the day identified in the foregoing schedule at or after the time on that day corresponding to the time that loading was completed and/or the vehicle was released, whichever is later.

In the event Carrier fails to comply with the above-specified schedule, line-haul transportation charges computed under the provisions of this contract shall be reduced by an increment of twenty percent (20%) for each day, or fraction thereof, of additional transit time, subject to the provision that under no circumstances shall such transportation charges be reduced more than eighty percent (80%) because of lateness in delivery.

(6) DELIVERY: Carrier shall tender shipments for delivery at its own convenience, subject to the provisions of paragraph (5) hereinabove; except that, if at the time of shipment Shipper directs Carrier to telephone the party to whom the shipment is consigned and obtain an appointment for making delivery, Carrier shall comply with this direction, the cost of such telephone call to be added to the freight charges.

If, in accordance with the schedule set forth in paragraph (5) hereinabove, a shipment would be available for delivery on a day, except Saturdays, Sundays and holidays, at a time such that free time for unloading as specified in paragraph (7) hereinbelow would expire before the close of business at the place of delivery, and Carrier is not permitted, through no fault of its own, to make delivery on that day, an additional charge of two hundred dollars ($200.00) shall be assessed for each day, or fraction thereof, that actual delivery is delayed.

(7) FREE TIME AND DETENTION: A total of three (3) hours shall be permitted for the loading of freight at origin, and a total of three (3) hours shall be permitted for the unloading of freight at destination, subject to no additional charge. An additional charge of ten dollars ($10.00) shall be assessed for each fifteen (15) minutes of time occupied for loading and/or unloading beyond the free time specified in this paragraph. The time for loading and the time for unloading shall be computed, for purposes of this paragraph, beginning at the time Carrier's vehicle actually arrives at the premises designated by Shipper for loading or unloading, or, in the event Carrier has been assigned and has

accepted a delivery appointment pursuant to paragraph (6) hereinabove, such appointment time, whichever is later, and shall end when loading or unloading is complete and the vehicle has been released to Carrier.

(8) TERMS, CONDITIONS, RATES AND CHARGES: Carrier shall perform the transportation described hereinabove pursuant to the terms and conditions of, and at the rates and charges specified in, [schedule of rates and charges]; except that the parties may mutually agree, prior to transportation of any shipment or lot of freight, to terms, conditions, rates and charges other than those specified in [schedule of rates and charges], such agreement to be in writing.

(9) PAYMENT OF FREIGHT CHARGES: Shipper agrees to pay all freight charges determined under the provisions of this contract for any transportation performed by Carrier pursuant to the provisions of this contract, except that shipments may be dispatched by Shipper, at its option, on a "collect" basis, with charges to be collected by Carrier from the consignee or from a third party. In the event that Shipper is not to be primarily responsible for freight charges, Shipper further agrees to guarantee payment of such charges should Carrier, despite the exercise of due diligence, be unable to effect collection of such charges, or any portion thereof, from the consignee or third party.

(10) CREDIT TERMS: Shipper agrees to pay all freight bills submitted by Carrier pertaining to transportation service furnished under the terms of this contract within fifteen (15) days from such submission. In the event that Shipper is in default of this provision, Shipper agrees to compensate Carrier in the amount of 2% of the net amount of each freight bill paid by Shipper within more than 15 days, but less than 30 days, following submission of such freight bill, and in the further amount of an additional 2% of the net amount of the freight bill for each 15 days or fraction thereof that the freight bill remains unpaid beyond the first 30 days following submission of that freight bill by Carrier.

(11) LOSS-AND-DAMAGE LIABILITY: Carrier agrees to assume the same liability for loss of, damage to, and/or delay of freight as is imposed on motor common carriers under the Interstate Commerce Act, as amended, and the common law of the United States of America; and Carrier further agrees that any claims for loss, damage or delay shall be governed by the provisions of the Interstate Commerce Act and the common law applicable to motor common carriers; *except that,* where loss, damage or delay is due in part to the negligence of Shipper,

its employees or agents, liability for such loss, damage or delay shall be pro-rated between Carrier and Shipper in proportion to the degree to which the aforesaid negligence of Shipper, its employees or agents contributed to the economic loss incurred.

(12) FORCE MAJEURE: The obligations of the parties under terms of this contract shall be temporarily suspended during any period in which either party is unable to comply with the provisions of this contract by reason of acts of God, acts of a public authority, acts of a public enemy, fire, flood, labor strike or disorder, civil commotion, closing of the public highways, or other contingencies, whether similar or dissimilar to those named, beyond the reasonable control of such party. In the event the obligations of the parties must be thus temporarily suspended, the time of such temporary suspension shall be excluded from the computation of time under the provisions of paragraph (17) hereinbelow; and the time periods specified in paragraph (17) shall be accordingly extended by the same amount of time as the period of the temporary suspension.

(13) NOTICES: All notices given pursuant to the terms and provisions of this contract shall be deemed sufficient if given in writing and sent to the party being noticed by first-class mail, with sufficient postage properly affixed, at the address designated hereinabove or such different address as may be specified by either party, in writing.

(14) SETTLEMENT OF DISPUTES: The parties herewith agree that any disputes between them arising out of the provisions hereof, and/or any transportation or service performed by Carrier pursuant to such provisions, shall be submitted to binding arbitration at the demand of either party. In the event the provisions of this paragraph are invoked, each party shall appoint one (1) individual, not in its employ or under contract, directly or indirectly, with it, to serve as an arbitrator; and the two arbitrators thus named shall name a third individual to serve as an arbitrator. The parties agree to abide by the majority decision of the three-member arbitration panel chosen as specified hereinabove, and expressly waive and abjure any rights they might otherwise possess to bring legal action against one another or otherwise dispute or appeal from the decision of such arbitrators.

(15) NON-DISCLOSURE: The parties agree that they shall not themselves, nor through agents, employees or by means of third parties, knowingly disclose the contents hereof, nor the contents of [schedule of rates and charges], to any other party whomsoever; and that they shall not knowingly permit, agree to or abet such disclosure by any

other party; except that Shipper may, in connection with *bona fide* auditing of its freight bills, disclose the contents hereof and/or the contents of [schedule of rates and charges] to the individual(s) or organization(s) engaged by it to perform such auditing subject to execution of an agreement by such individual(s) or organization(s) that he, she, it or they, as the case may be, shall hold such information in confidence and shall not disclose such information to any person or organization not a party hereto.

(16) ENTIRE AGREEMENT: This contract constitutes the entire agreement between the parties, and may be modified only by mutual agreement of the parties as evidenced in writing.

(17) EFFECTIVENESS AND TERM: This contract shall take effect on the date it is executed by the parties, as set forth hereinbelow, and shall remain in effect for a period of six (6) months thereafter. Upon expiration of the aforesaid six-month period, this contract shall continue in effect thereafter from month to month, subject to termination by either party upon thirty (30) days' written notice to the other. In the event either party gives such notice of termination, all of the obligations and duties set forth herein shall continue in effect until after expiration of the aforesaid 30-day notice period, unless otherwise agreed to in writing by the parties. IN WITNESS WHEREOF, the parties hereby cause this agreement to be executed on this ____________ day of __________, 19____.

For Carrier ____________________ For Shipper ____________________

Witnesses:

____________________ ____________________

9

THE MANY FACETS OF COMPETITIVE BIDDING

You won't find much in most basic marketing or procurement texts about competitive bidding. It's not often discussed in any depth in business periodicals. Few companies—suppliers or purchasers—have resident experts versed in all its ramifications.

But competitive bidding has attained a significant place in certain economic sectors as a powerful (albeit limited) tool in the business negotiations process. And the post-deregulation transportation industry is one of those sectors.

Broadly speaking, competitive bidding is a kind of auction in which the buyer specifies what he wants to purchase and interested vendors respond by stating the price at which they're willing to sell what's been described. All bids must be submitted simultaneously without, of course, any bidder knowing what anyone else has bid (for that reason it's also sometimes known as "blind bidding"); there are generally no opportunities for bids to be revised; and, generally speaking, low bid takes the traffic.

Government is the heaviest user of competitive bidding. At all levels (local, state and federal), governmental agencies are major consumers of both goods and services. For obvious economic reasons they constantly strive, like other consumers, to purchase at the lowest price; and for political reasons they must assiduously avoid any hint of favoritism or discrimination in their buying. Competitive bidding gives them a way of achieving both objectives at once, to the degree that in many instances its use is mandated by statute and/or regulation.*

Organizations in the private sector have found that, although they of course lack the political incentive to make use of competitive bidding, the economic advantages of this approach to vendor negotiations may alone justify it in some circumstances. While by no means the most popular negotiating technique, competitive bidding has as a result won a certain amount of acceptance in the marketplace.

Unfortunately, in large part because of widely publicized reports on government procurement practices, competitive bidding has developed something of a bad reputation. Interestingly, the two principal criticisms are in large part mutually contradictory.

At one end of the spectrum, the lack of esteem in which competitive bidding is often held is epitomized by the oft-repeated wisecrack about military procurement: "Do you realize we're defending our country with weapons built by the lowest bidder?" The implication, of course, is that least-costly is necessarily equatable with lowest-quality; and highly publicized tales of helicopters breaking down in the Iranian desert and tanks that can't travel more than 20 miles without needing shop time do nothing to change this perception.

Yet at the other extreme there are serious doubts that the process is actually effective in minimizing procurement costs. Oft-reported stories of coffee pots, toilet seats, even simple hammers and wrenches costing hundreds of dollars have caused the Defense Department considerable embarrassment. If competitive bidding is so efficient a procurement method, it is asked, why these ridiculously high prices?

Such problems do not, however, truly have their roots in the

*The discussion of government uses of competitive bidding, here and subsequently, also pertains in some cases to private-sector firms working under government contracts. In some cases such firms are obliged to use government-mandated competitive bidding procedures in any sub-contract work they commission.

competitive bidding process itself; rather, they have to do with its misuse. As Cassius says in Shakespeare's *Julius Caesar,* "The fault, dear Brutus, lies not in our stars but in ourselves. . . ." It's up to the buyer to make competitive bidding work; if it fails to produce satisfactory results, he should blame himself, not the process.

The Request for Proposal

The key to competitive bidding is the bid solicitation prepared by the buyer. It is known by various names—"invitation for bid," "solicitation for offer," "bid request," and others. The designation employed here is one commonly used by many government procurement offices—"request for proposal" (RFP).

Developing a request for proposal is significantly different from preparing for other types of business negotiations. The normal negotiation process is one of give and take, in which the parties work gradually toward agreement. Any initial proposals by either side are merely starting points for the bargaining that will follow, and both perceive them that way. For that reason it's generally considered good negotiating tactics to ask initially for more than you expect to get, leaving yourself latitude for some concessions as the negotiations progress (*see Chapters 5 and 7).*

In competitive bidding, however, the buyer's initial proposal—the RFP—is intended to form, as written, the basis for the ultimate agreement; the process affords no opportunity for give-and-take bargaining that might change any of the particulars. Thus, a much more exacting approach to preparation of the RFP is indicated than if only an initial negotiating position were being developed.

There are six basic elements to the RFP:

(1) The specifications, or "specs"—the buyer's description of what he wants to purchase, which he is saying are not negotiable. This element will comprise the bulk of the RFP, as it must include considerable detail as to *precisely* what is wanted.

(2) Scope of the proposal—how large a purchase is planned? With respect to transportation service, this concerns principally

the volume of traffic that is involved, the span of time covered, etc.

(3) The bid element(s)—that is, the basis on which bids are to be submitted. Most commonly, price will be the only bid element, with bidders being asked merely to quote what they propose to charge for the required goods or services. Occasionally, however, one or more other factors may be identified as bid elements in addition (or, very rarely, instead), with bidders given latitude in these areas as well.

(4) Bid format requirements. The buyer may insist that bids follow a standardized format, be attested to by the signature of a responsible corporate officer of the bidder, ask for such other documentation as a current financial statement, description of how the bidder means to comply with the RFP standards, resumés of the personnel who would be involved, etc.

(5) Criteria for bid evaluation. Where multiple bid elements are involved, it's obvious that the RFP must identify how much weight is to be accorded to each. Even where price is the only variable, the buyer may want to include other factors in his evaluation—for example, the bidders' financial condition, past records of dependability, etc.

(6) Deadline for the submission of bids—date and, in most cases, time of day.

Failures of the competitive bidding process may almost invariably be ascribed to the buyer's having paid inadequate attention to one or more of these six areas.

The Specifications

"Be careful what you wish for," goes an old saying. "You might get it." Where competitive bidding is involved, this is more than a possibility; it's a virtual certainty, because the buyer has taken on the entire responsibility for defining his needs and soliciting vendors who will engage to meet precisely those needs.

A transportation RFP begins with a description of the traffic the shipper wants to move. First, what is (are) the commodity(ies) involved? Commodity identification should be as specific as possible, not just generalized descriptions taken from existing Classifications or tariffs and/or Standard Transportation Commodity Code (STCC) numbers (although these may be incorporated in the description as well). Information concerning such key elements as unit weight, density, value, form of packaging, fragility, protective-service needs, hazardous nature (under U.S. Department of Transportation and/or international rules), etc., should normally be offered where pertinent.

Where multiple commodities are involved—especially if the shipper is soliciting bids for a single rate scale that will apply uniformly to all his traffic on a "freight all kinds" type of basis—it will be important to indicate the proportionate volume of each that is expected to move. Necessarily this entails a certain amount of guesswork, since it's of course impossible to know for certain exactly what the future may bring. But the shipper assuredly will have developed his own forecasts for procurement, sales/marketing, etc., purposes, and should be prepared to share this information with the carriers whose bids are being solicited.*

Although few carriers are likely to take advantage of it, it may also be a useful to give prospective bidders a chance to examine both the goods themselves and their protective packaging—especially if less-than-truckload movements, where carriers will be obliged to platform the shipments in transit, are involved. (This will also let the carriers take a look at the facility(ies) where they would be expected to make pickups and/or deliveries, another possible aid to them in making their bids.)

Some shippers ensure this by commencing the competitive

*Since such forecasts are often closely guarded competitive secrets, a dilemma can arise: If the shipper reveals the information fully and indiscriminately through an RFP, it will surely get back to his competitors; but if he withholds all forecast data, he's giving the carriers insufficient information on which to submit rationally formulated bids. One solution is to limit the amount of forecast information included in the RFP, such as by breaking out traffic volumes only according to differences in class rating or other broad-brush characteristics. In other cases shippers will make this information available only to serious bidders who request it, requiring them to sign nondisclosure agreements first.

bidding process with a formal presentation—a meeting to which all bidders are invited where the specifications of the RFP are described orally, and carriers may ask any questions, before the documentation is made available to bidders. This certainly maximizes the bidders' opportunity to get information that may help them in bidding. At the same time, since prospective bidders are obliged to incur the time (and probably travel expenses) necessary to attend such meetings in only the speculative hope that they may win traffic, this approach may discourage carriers who would otherwise submit bids. Both factors must be carefully weighed before a meeting of this nature is scheduled as part of the competitive bidding process.

The shipper must also define the geographic considerations pertinent to the traffic—the origin(s) and destination(s) between which he wants service. If possible, each separate point-to-point routing, again with approximate percentages of the traffic moving over each, should be named. If geographic diversity is too great for this to be feasible, there will normally at least be a limited number of points at one end or the other of the movements in question (origins as to outbound traffic, destinations as to inbound) that can be thus identified; and groupings, regional delimiters, etc., may be employed to identify where the traffic will move to or from.

To this point preparation of the RFP has been fairly straightforward; the consist and geographic scope of the traffic is (or should be) well within the clear knowledge of any shipper. (If future movements are truly so uncertain that specificity in these areas isn't possible, the shipper is ill-advised to be using competitive bidding at all.) The next step, however, will be appreciably more difficult: The shipper must identify exactly what level of service he wants.

The difficulty here lies in the fact that shippers are not experts in transport operations, whereas quality-of-service matters are, for the most part, operations-oriented. For that reason the shipper must tread carefully indeed as he approaches this area.

The place to start is with past service, and the terms and

standards under which it has been rendered.* Tariff or contract rules and standards under which traffic has moved previously will provide the basic framework for service specifications, and should be researched in considerable detail as this portion of the RFP is prepared.

That doesn't necessarily mean such existing tariff/contract standards should be followed slavishly; they must obviously be viewed in light of the shipper's past experience. If the shipper has been reasonably satisfied with the service he has received, he may choose simply to adopt those standards wholesale (making certain, of course, that they fully cover the service that's actually been provided). If he's dissatisfied in one or more respects, on the other hand, appropriate changes in the existing standards should be incorporated in the RFP.

Specifically, matters that must be taken into consideration are:

- *Desired transit times.* If "reasonable dispatch"—the basic, rather vaguely defined standard of the law that applies to common carriers—is sufficient, that may be specified; but in most cases it will be preferable to spell things out in greater detail.
- *Accessorial services.* These are anything apart from the actual line-haul transportation. Such services as stopping in transit for partial loading/unloading, multiple pickup/delivery operations, "peddle-run" deliveries, sorting and/or segregation of freight, exclusive-use-of-vehicle service, in-transit storage, diversion/reconsignment, special in-transit security requirements, Customs in-bond movement, provision of hoses and other connectors for loading/unloading bulk freight, etc., come readily to mind; others (and the list is all but endless) will occur to shippers in conjunction with their particular traffic.

*An inexperienced shipper—a company that, for example, is just starting up in business and has no background of past traffic behind it—cannot, of course, reference a history it hasn't got; and for that reason such a shipper should probably avoid competitive bidding on the initial movements of this traffic; the risks of misdefining the level of service it will need are simply too great.

• *Carrier equipment.* What type of equipment is wanted? When, where and under what circumstances is empty equipment to be positioned for loading (and/ or loaded cars for unloading)? Is the carrier to provide empty equipment suitable for loading on a predetermined schedule, or on demand (and, if so, with what advance notice)? What effect will carrier failures to meet these standards have in terms of (1) volume commitments, and (2) the shipper's possible need to pay higher charges to move freight via other carriers or modes because of equipment unavailability?

• *Free time and demurrage/detention.* How much free time for loading/unloading will be allowed? Are allowances to be made for "bunching" (when so much freight arrives at once, due to vagaries in carrier transit schedules, that it overtaxes the capacity of the receiving facility to handle it without delays)? On what basis will demurrage/detention charges be assessed? If averaging agreements (common in the rail industry) are to be used, will demurrage debits/credits on the RFP freight be handled separately or will they be lumped with other service?

• *Billing and payment terms.* The amount of credit extended to the shipper, and any discounts/penalties that may apply to ensure compliance with credit terms, will be pertinent factors.

• *Loss-and-damage liability.* The RFP should identify clearly the extent to which the shipper will look to the carrier for compensation should goods be lost, damaged or delayed while in his custody. Alternatives range from a negligence-only standard at the low end of the scale (that is, the carrier is liable only if he was provably negligent) to full common-carrier liability "as a virtual insuror" at the high end, with a full spectrum of options in between.

• *Loading/unloading.* This obviously has application primarily for motor carriage, although in some instance it may also be a consideration in connection with service via rail or water. It will have a significant effect on carrier costs (and hence bids), and should accordingly be specified where it is a pertinent consideration.

• *Co-loading.* Again, this applies mainly to motor carriage, although in certain circumstances it may have a bearing on air service. Left to their own devices, carriers will almost certainly seek to fill any voids in trailers or containers that have leftover cubic capacity. If the shipper doesn't want this done (it may result in transit delays, and also tends to increase the likelihood of loss and damage), that should be specified in the RFP.

The foregoing does not purport to be a comprehensive listing. Given the wide diversity of traffic and transportation needs in an economy as variegated as that of the modern-day United States, such a list would be almost impossible to assemble, and would have to include many obscure matters of only the most limited application. The items identified here are intended only as guidelines for the shipper in compiling the details of his own RFP, and certainly should not be regarded as necessarily all-inclusive.

Beyond any question, the single biggest reason for competitive bidding failures is inadequacy of the specifications. This is what underlies most of the government procurement problems that have been so widely reported in the news media.

In order to use the competitive bidding process effectively, a shipper is going to have to commit significant volumes of traffic to the successful bidder(s) (*see below).* Accordingly, he is going to *de facto* make himself a "captive shipper" to one or a few carriers, foregoing the freedom of the competitive marketplace. And the carrier(s) to whom he has placed his traffic in thrall are going to be offering, as the inevitable result of the bid process, precisely the level of service specified—no more and no less.

If that service is unsatisfactory—if the shipper has forgotten about needed elements, or misstated them, or has otherwise failed to define his requirements accurately and clearly—there are only two options, both unsatisfactory. Either the shipper must endure the sub-par service for the duration of his commitment, or he must seek to negotiate the changes he needs *under the captive-market conditions he himself has brought about*—a negotiation in which, obviously, he cannot expect to do well.

At the same time, it is likewise a mistake to overdo the specifications process—an extreme to which many RFP preparers in all economic sectors resort out of fear of under-specifying. This is a common problem in government procurement offices. To offer one example taken more or less at random, Defense Department specifications for fruitcake to be served at Christmastime in military mess facilities occupy 12 densely printed pages; by comparison, a fully detailed fruitcake recipe requires only half a page or so in most cookbooks.

Such overspecification obviously discourages bids; the carrier who must digest an RFP running to 50 or 100 pages or more will obviously be more inclined to forget about bidding than if he had a less voluminously detailed document to deal with. An overlong RFP may also generate less, not more clarity simply by its very wordiness. And it also tends to increase the likelihood of error as the RFP preparer becomes so immersed in details he loses sight of the main objective.*

Completeness and accuracy are of unquestionable importance in an RFP; but so is brevity. In its way overspecification of the service a shipper requires can be just as detrimental to the process as can underspecification.

Traffic Volume

Next in order is the volume of anticipated traffic covered by the RFP. This need not, and should not, include every pound of freight expected to move; it should, rather, be a *minimum* volume which the shipper undertakes to guarantee to the successful bidder(s).

For the competitive bidding process to be really successful, a significant volume commitment must be offered. By choosing the

---

*In this context, a perhaps-apocryphal (and perhaps not!) story is told of a Defense Department procurement effort involving a new weapons system. Literally thousands of pages of the RFP were devoted to specifying each detail of the system, down to the strengths of each component, the size fasteners to be used, etc., etc. The contract was let, and after a time military observers were invited to see the prototype. They inspected it carefully, found each detail in full compliance with the specifications, and at last asked for a demonstration. "Oh, it doesn't *do* anything," said the contractor in surprise. "You didn't say it had to."

competitive bidding process in the first place, the shipper has opted for a negotiating approach that is extremely volume-sensitive by nature. That is, carriers will look closely at the volume of traffic at stake before deciding (a) whether to respond at all, and (b) if they do respond, how low they should bid. To many carriers the lure of substantial traffic volumes is a powerful one; thus, the more volume is offered, the more carriers are likely to be seriously interested.

In fact, unless there's a significant volume of traffic involved the bid process becomes a sham. Following deregulation, some shippers adopted the practice of telephoning various carriers for "bids" before moving so much as a single shipment, deluding themselves (and often their bosses) into believing they were thereby engaging in competitive bidding.

Especially where common rail, motor or domestic water carriage, or surface freight forwarding, are involved, this procedure is either (1) useless, or (2) dangerous. As of this writing, these carriers were obliged by law to charge, and their shippers to pay, the rates and charges set forth in their tariffs; thus, it is both more accurate and (generally) quicker to check such a carrier's rate in its tariff than to make a telephone inquiry. The element of danger enters the picture when carriers take the shipper's inquiry to be an invitation to quote non-tariff (illegal) rates; numerous shippers have found to their dismay that such quotations don't protect them when the carriers seek, often long after the fact, to collect tariff-based undercharges (*see Chapter 3).*

Even where traffic not subject to tariff-publication requirements is involved, such casual polling for shipment-by-shipment rate quotations will generally have more cosmetic than real value; it may *appear* that the shipper is getting the lowest rate available (since he's accepting the lowest among several quotes), but in fact the strong likelihood is that he could do much better by other means. Without significant traffic volume at stake, carriers will in most cases not take the trouble to prepare special quotations; they'll simply check the "standard" rate in their tariffs, rate circulars, schedules, etc., and quote it.* In effect, thus, the shipper is merely checking out existing prices on the "spot" market, much as a consumer would shop for household purchases, rather than engaging in anything that could be called serious competitive bidding.

Those involved in other forms of business procurement have long abjured this simplistic approach. A company seeking to purchase office supplies, for example, does not telephone local discount retailers to learn which has the best price on a given day; rather, it lumps its purchases and buys from the same wholesalers who service the retailers, eliminating the retail markup that even the deepest discounter must take. Spot-market purchases are reserved for emergency situations when there's an unforeseen immediate need for small quantities; and higher prices are taken for granted. It's elementary that the same approach is equally applicable to purchasers of transportation services.

If one *must* buy on the spot market on a particular occasion, the telephone-poll approach may have some value. It *does* afford an opportunity to do some rate comparisons; and in some cases, where tariff-publication restrictions don't intrude, the shipper may be fortunate enough to contact a carrier willing to offer a sharp on-the-spot rate reduction—one, perhaps, with uncommitted empty equipment on hand that he's eager to load. But as a regular procurement practice this is virtually certain to be ineffective as a means of minimizing transportation costs; and the shipper who calls this sort of thing "competitive bidding" is merely using the nomenclature to disguise from himself the need for a more coherent negotiating strategy.

In sum, competitive bidding, to be truly effective as a negotiating tactic, *must* be supported by sufficient volume to attract

---

*The accuracy of such quotations is, moreover, by no means assured—and, as already discussed, oral quotations are not binding by law where tariff-publishing requirements exist, and may be difficult to enforce even against carriers *not* subject to such requirements if the rate circular or schedule specifies otherwise.

the carriers whose bids are solicited. Without such a level of volume, the process is useless, or worse.

At the same time, there's no point in the shipper's carrying his description of volume in the RFP beyond reality by referring to more traffic than he's in fact probably going to have available. A reasonable minimum volume should be estimated, and any enforceable commitments should allow a margin of safety below this level (to leave room for unexpected economic downturns and the like). Under no circumstances should the shipper incur a serious risk of falling short of his volume commitments; should that ever happen, he should immediately begin an exhaustive reexamination of his forecasting techniques, which have obviously gone seriously awry.

Another question that must be answered is the length of time over which the shipper wishes to make his commitment. Again, the period must be long enough to encourage serious responses from carriers; but it should also be short enough not to lock the shipper in to an arrangement that the passage of time may render uncompetitively expensive. To some degree this will depend on the volume of traffic involved, but six months will be a good base timeline, and in today's volatile transportation market it will usually be unwise to extend commitments under a competitive-bidding award for more than a year. At the same time, there is no reason any agreement should not be extendible, at the will of the parties, for the indefinite future under an "evergreen" or similar proviso.*

The Bid Element(s)

When it comes to the bid element(s) of the RFP, price is, as already noted, usually the only one. That is, prospective bidders are given the definition of the service required and the volume

---

*In some specialized types of transportation agreements, of course, the calendar may not be the measure of duration. For example, competitive bidding may be used for transportation to job sites during the life of a construction project, the term of which may be uncertain.

commitment, and are asked to specify a rate, or rate scale, under which the service will be provided. But there is one factor that need be first considered; and there are some other options that may be explored in appropriate circumstances.

First, the RFP should define the form of carriage that is desired—that is, common or contract. It is, of course, possible to leave this question open-ended, allowing carriers to submit bids on either basis; but in most cases the shipper will have a predetermined preference, which should be reflected in the RFP.

In these deregulated days, the main difference between the two forms is the manner in which the ultimate agreement will be expressed—whether it will be in a tariff, subject to public scrutiny and open to unilateral change by the carrier, or in a contract that may be held secret and is bilaterally binding.

The advantages of contract carriage are fairly succinctly capsulized in the foregoing: Rates and charges need not be revealed to others (competitors, etc.); and the carrier, having agreed to a particular deal, is legally bound to keep it. But sauce for the goose is, in this regard, also sauce for the gander; just as the carrier is obliged to keep *his* agreement, so is the shipper. Where common carriage is involved the shipper's commitment will normally be expressed in terms of discounts off existing rate levels, so that, should volume fall short of the agreed level, the shipper will incur no worse damage than having to pay normal (undiscounted) tariff rates. A contractual arrangement, by contrast, may expose the shipper to deficit-volume provisions that oblige him to pay for traffic that never moved.

Because of this mutually non-binding nature of the commitments—the carrier being free to change rates at its sole option, the shipper to curtail traffic without severe penalty—competitive bidding will not often prove a useful approach to common carriage, since either party can *de facto* abrogate their agreement at will (thereby rendering all of the time and effort that has gone into the competitive bidding process useless).* It is

*However, the discussion below notes certain circumstances under which this will not be the case—where competitive bidding does work well in common carriage.

best suited, in most instances, to the establishment of contractual relationships that entail meaningful and legally binding commitments from both parties.

And although price is usually the only object in competitive bidding, this won't always be the case. It may be that, in addition to the rate(s) or rate scale(s) requested by the shipper, certain other questions are left to carrier discretion.

Most commonly this will be done with respect to other dollar-oriented issues, such as accessorial charges, detention/demurrage charges, and the like—charges, that is, which are normally stated separately in carrier tariffs and contractual rate schedules, inasmuch as they apply to services or privileges that are only sometimes wanted. In such cases the evaluation process usually entails modeling of a few representative shipments (based on historic traffic and movement data) that reflect the expected average incidence of these ancillary charges.

But the shipper may also wish to give prospective bidders latitude in certain other regards, too. The degree of carrier loss-and-damage liability, equipment specifications, billing/payment terms, whether loaded vehicles will move under seal (precluding co-loading) or not, etc., are possible areas where the RFP may leave things up to each individual bidder.

If latitude is allowed to bidders in such ancillary areas, it *must* be rigorously controlled. Failure to do so is what often accounts for those costly coffee pots, tools and the like, for which government has been, justly, publicly excoriated. What has happened, quite evidently, is that a weapons-system bidder has "lowballed" his price for the system as a whole—the primary bid element—with the intention of making up lost profits by means of extraordinarily high pricing in secondary areas which he quite accurately anticipates will be largely overlooked or ignored in the bid-evaluation process. The same fate can readily befall incautious transportation shippers who fail to observe that a bidder's low line-haul rates are accompanied by vastly inflated accessorial charges and the like.

Moreover, the more variables are allowed in bid responses, the more complex and difficult will be the evaluation process. It's obviously necessary to compile the results in such a fashion that different bids, taking different approaches to the different bid elements, may be compared with one another, and the task quickly becomes onerous when multiple bid elements are involved. The most usual approach is accordingly to allow only one or two bid elements, the primary criterion being price.

During the early days of deregulation a few large shippers went to the extreme of allowing no bid elements at all. They developed their own "suggested" tariffs, which they then "encouraged" common carriers, through proposals very similar to the RFP, to publish with the ICC and/or state regulatory agencies. Carriers that acceded to this demand were promised a fixed share of the shipper's traffic (often on a first-come-first-served basis); carriers that didn't had to settle for the dregs of the traffic, if any at all.

Such an approach loses, obviously, many of the benefits of competitive bidding; no carrier is at all likely to offer the shipper any better deal than he's specified. It does, however, have the advantage of allowing the shipper (provided his proposed rates are high enough to be attractive) a reasonable selection of carriers, all of which will have identical rates, rules, service levels, etc. In some instances this uniformity may be highly desirable to a shipper, warranting the reduced opportunity to select among competing bidders.

Using this approach, the shipper actually prepares a complete model of the final documentation he wants for both submission of "bids" and the final shipper-carrier "agreement," leaving blanks only for insertion of each "bidder's" name, address and alphanumeric tariff designation. "Bids" are then submitted in the form of proof that the tariff has indeed been filed with the regulatory agency.

Bid Format

Even shippers who take a more conventional approach to competitive bidding, leaving one or more elements open for bidders to determine themselves, should insist that bids be submitted in similarly standardized form.

Consider the shipper's problem on the day bids are opened and must be evaluated. First he must go over each to make certain all the service standards he has specified are included. Then he must analyze the bid elements he has left for carriers to complete, weighing them one against another. Under the best of circumstances this process will take appreciable time to accomplish if any substantial number of bids have been received; and it should be obvious that a fixed bid format will greatly expedite the task.

Furthermore, reliance on a standardized format will avoid potential problems due to inconsistencies of terminology employed by different bidders. If he's done his work properly, the preparer of the RFP has put considerable thought into the phrasing of his specifications. How, then, is he to deal with bids that phrase the same specifications in other language; how can he be sure that it's merely the words, and not the meanings as well, that have been changed?* It seems both excessive and unnecessary to require that evaluators spend long hours poring over the various bids to make sure each carries the same meaning, when the whole problem can be readily avoided by obliging each bidder to use a standardized bid format.

The same considerations make it desirable that the bid format be extremely close, if not identical, to the form the shipper's final agreement with the successful bidder will take. That is, if common carriage is involved bids should be submitted in tariff form; if contract carriage, a model-form contract should be employed. In both cases, the format specified by the RFP should incorporate all of the RFP specifications, with carriers inserting only the bid elements in their submissions.

There is, it should be noted, one other factor that mitigates in favor of as much standardization of the bid format as possible. For the most part transportation carriers—especially smaller carriers—are largely unfamiliar with the discipline of com-

*The problems of "plain English," as distinguished from more technical terminology, have already been discussed; see Chapter 8.

petitive bidding. By greatly simplifying the process of submitting a bid, a standardized bid form encourages responses from carriers who might otherwise be uncertain just how to proceed.

By now the perceptive reader will have noted several references to the desirability of encouraging carriers to submit bids. Competitive bidding is only useful as an economic negotiating tool to the extent it engenders responses from a representative segment of the market. Thus, it is axiomatic that a bidding process that develops but a smattering of responses is not likely to be an effective one.

This axiom has two important corollaries:

First, if the market itself is small—if, for example, a highly specialized type of service is involved, which few carriers are equipped to provide—the process should be eschewed in favor of head-to-head negotiations with one or more of those carriers. At root, competitive bidding is no more than a means of sampling a market too large to practicably investigate in detail; if the market does not meet this criterion, the process pointlessly sacrifices the possible benefits that may be available through one-on-one negotiations.

Second, the same consideration holds true if the shipper has elected for reasons of his own to narrow the selection to a very small number of carriers, who alone will be allowed to bid. Rather than using the single-offer approach of competitive bidding, the shipper, again, will generally do better by negotiating independently with each such carrier.

By its very nature, the power of competitive bidding lies in its ability to generate a wide array of responses without an excessive commitment of time and effort on the shipper's part. Accordingly, every effort should be made to encourage carriers to bid, by minimizing the time and effort *they* must dedicate to the process and by enhancing the rewards to make bidding appear as attractive as possible.

The extent to which carriers are obliged to submit other information along with their bids will to a considerable extent depend on how the shipper has elected to circulate the RFP. Some competitive bidding is on an invitation-only basis; that is, the RFP is offered only to a select list of vendors, and only their bids will be considered. In such cases the shipper will obviously have developed information about prospective bidders before he invites them to take part, and will probably need little or no additional information during the bidding process.

An alternative approach is to advertise availability of the RFP far and wide, offering any vendor who wishes the chance to bid. The classified advertising sections of trade journals, form letters to carrier associations, traffic clubs, etc., or even to carriers themselves, may be used for this purpose. In these circumstances the shipper will be expecting to receive bids from carriers about which he may know nothing at all. Since he's contemplating a major commitment to the successful bidder(s), he should, therefore, insist that carriers include with their bids appropriate additional information. Financial reports, balance-sheet information, details about equipment and/or personnel, OS&D (over, short and damaged) records and the like may be required to demonstrate the carriers' qualifications to handle the traffic.

The Basis of Evaluation

There are two purposes to specifying in the RFP the basis on which bids will be evaluated. The first and most obvious is to ensure that carriers zero in on the areas of greatest importance to the shipper in preparing their bids, and don't waste both their and the shipper's time with bids that don't really address the principal points. Knowing where the shipper places his priorities will help carriers better allocate their resources to meet those priorities.

More subtly, inclusion of expressly stated bid-evaluation criteria sends a message to prospective bidders that their submissions will be looked at fairly and objectively. This is especially important when bids are being sought widely through advertising methods. Often a carrier who has done little or no business with a particular shipper will be reluctant to dedicate much time or effort to preparing a bid, in the belief that his competitors probably have the "inside track." If the RFP spells out clearly the criteria under which bids will be evaluated, however, a good deal of this reluctance can be overcome.

The RFP statement on this point should include (a) clear identification of every factor the shipper will be taking into account in evaluating the bids, and (b) the proportionate weight to be assigned to each. Generally this will be stated in percentage terms; for example, "Price 60%, financial stability 20%, equipment 10%, experience of personnel 10%", or something of this nature.

This can become quite complex where a fairly broad grouping of movements, as to which carriers are asked to submit rate scales (rather than a single rate), is involved. How does the shipper propose to compare one bid that's fairly high for short-haul movements but tapers considerably for greater distances, and another that's lower for short trips but has less taper at the long-haul end? How will he evaluate a bid with low LTL and high truckload rates against one that offers the reverse? If the shipper is at all serious about competitive bidding, he will certainly have to develop some objective basis on which to make these comparisons (proportionate weighting, application to representative shipments, etc.); and he will certainly improve his likelihood of receiving the lowest bids if he reveals this information to the carriers in the RFP.

Some shippers seek to simplify this process, where widely diversified traffic is involved, by evaluating discreet segments of that traffic separately. Thus, for example, there may be one evaluation for short-haul traffic and another for traffic in long-haul markets, or one for LTL and another for TL, etc., with each award being made independently (and carriers, of course, encouraged to bid only on those segments they want to handle). In actuality, what the shipper has done is to consolidate multiple independent RFP's into a single document for administrative convenience, with only the evaluations being kept separate. Once again, if this approach is to be taken it should be specified in the RFP.

One important aspect of the evaluation process, however,

should generally *not* be revealed in the RFP. At regular auctions of goods it is common for sellers to establish a "reserve" price—a price below which an item is not to be sold. Thus, if bids don't match or exceed the reserve level, the item is withdrawn. (Where bidding is light, some professional auctioneers will pluck phantom bids out of thin air during the course of the auction itself, in the hope of pushing a real bidder up to or above the reserve.)

The same tactic is also appropriate for a competitive bidding "auction." Before starting the bid process, it's a good idea for the shipper to perform at least some rudimentary market research to determine the level of rates presently available for the service he wants. This then becomes the "reserve"; if no bid betters that level, none will be accepted.

This section of the RFP should also detail just how the shipper proposes to allocate the traffic. If it's a winner-take-all situation, with the low bidder to receive 100% of the subject traffic, that should be specified. Many shippers, however, prefer to copper their bets somewhat by accepting more than one bid and spreading the traffic among several carriers on a proportionate basis: for example, 50% to the low bidder, 25% to the second-low, 15% to the third-low, and 10% to the fourth-low. (This approach also helps encourage more bidding, since carriers can hope to enjoy at least some of the traffic even if they're underbid.)

Yet another tactic used by some shippers is to employ the bidding process merely to "qualify" a fixed number of carriers for undetermined amounts of future traffic. For example, a shipper might specify that the 10 lowest bidders will go on a list to be notified when and as traffic becomes available.

This is one way in which competitive bidding may be of great value to the shipper that wants to retain the flexibility and price-competitiveness of common carriage while keeping the administrative workload of its Traffic Department down to manageable proportions. Especially if the mode being used is motor carriage (although to some degree for other modes as well), the shipper may have a very large number of carriers available to handle its traffic.

As a practical matter, however, it will be virtually impossible to keep tabs on the rates of every carrier—especially since, with deregulation, rates can change literally overnight. To make the task administratively feasible, the shipper must find some way of restricting the number of carriers to be considered; and this approach to competitive bidding, with low bidders being placed on a "qualified" list for a defined span of time and no other carriers eligible to participate in the traffic during that period, limits the number of carriers whose rates must be scrutinized while ensuring that their rates will be among the lowest available.

Another sometimes-used variant on this theme, however, is usually less effective. Some shippers will instead use competitive bidding to create lists of carriers with whom they later plan to negotiate further on a one-to-one basis. In such circumstances carriers will be torn between two motivations; they want to submit bids low enough to ensure that they'll be among those with whom negotiations are conducted, but they want to leave enough leeway for the further concessions they expect the subsequent negotiations will entail. Faced with this kind of conflict, a carrier can easily misjudge and submit a bid too high to warrant its inclusion on the list—even though, in negotiations, that carrier was prepared to go lower than any other bidder.

In sum, it will generally be better to keep the bidding process "pure" by requiring that carriers submit their best offer in the first place. If anything else is involved, the competitive bidding process will probably be the wrong negotiating tool to use for the job at hand.

The Deadline

Finally, a firm deadline for the submission of bids must be fixed. Generally, this is phrased in terms of the date and time by which bids must be *in the hands of the shipper*—that is, be physically at a certain address. The alternative of allowing postmark dates has the obvious drawback of leaving matters in the uncertain hands of the U.S. Postal Service, which may normally deliver the mail promptly but which has been known to experience delays of considerable duration.

The shipper should gauge his "lead time"—that is, the interval between his initial offering of the RFP and the deadline for bids—carefully. He wants to leave enough time for carriers to carefully consider the terms of the RFP, but not so much that they're encouraged to procrastinate in preparing their bids. Depending on the circumstances (in particular, how widely the shipper wishes to disseminate the RFP), that time span can be anywhere between a few days and perhaps six weeks; the median, if such a thing can truly be said to exist, is probably about two weeks.

The RFP should also specify that (1) no bids will be opened until the deadline has passed, even if they're submitted early; and (2) any bidder may rescind or amend his bid up to the deadline. Where government is involved, such provisions are mandatory to ensure that the competitive bidding process won't be sabotaged or subverted; in the private sector, where competitive bidding has less stringently binding force (*see below),* such pledges simply help communicate to vendors the integrity of the process and thereby encourage them to bid.

Responding to the RFP

Few carriers—indeed, few vendors in any economic sector—really like competitive bidding. They feel, with some justification, that they are being pressured to trim prices to the bone in an effort to overcome unseen competitors. Much preferable, from their standpoint, would be head-to-head negotiations where they might at least hope to get some inkling of what kind of competition they're facing before making concessions that may be competitively unnecessary.

This, of course, is precisely the frame of mind the shipper is hoping to engender. At least up to a point, the shipper wants to encourage carriers to fear phantasmagorical competitors, to worry about what the "other guy" might do without any clear indication of what that amorphous party is *actually* doing.

There are two ways for carriers to avoid falling victim to this form of psychological pressure. The first and simplest, of course, is simply to refuse to respond to competitive bidding invitations at all, as a matter of policy. Some carriers adopt this as a firm policy, and their reasons are not altogether incomprehensible.

The alternative—admittedly more difficult, but certainly more profitable in the long term—is for the carrier to develop, and steadfastly stick to, marketing strategies that reflect what *he* wants, without overconcentration on what the competition may or may not be doing. That doesn't, of course, mean ignoring competitive circumstances; no carrier can prosper if his rates are invariably higher than those of his competitors. But it *does* mean developing and maintaining a clear idea of where one's individual "bottom line" lies, and ensuring that all negotiating proposals, no matter how developed, don't violate that measure.

In short, in responding to any RFP the carrier should make certain that his bid, if accepted, will enhance his *overall* profitability. No matter what he believes his competitors may be doing, the carrier who intentionally "lowballs" a bid—or worse, the carrier who bids based solely on his assessment of the competition without reference to his own costs or strategic objectives—will emerge as a loser in the long run even if his bid is successful.

All of the foregoing should really go without saying; nothing could be more obvious than the carrier's need to make every significant increment of traffic contribute to, and not detract from, his overall profit picture. In the real world, however, it is relatively easy to be caught up in what is sometimes called the "auction mentality"—the notion that any bidding process amounts merely to a game that is to be "won" at all costs, without reference to extraneous circumstances.

How much one should gamble—as any frequenter of the tables of Las Vegas or Atlantic City, the horsetracks, etc., knows—depends on how much one is prepared to lose; no-one, obviously, should risk more on a roll of the dice or turn of the cards than he can afford should the fates decree against him. In competitive bidding the only difference is that the situation is inverted; the proper question to ask is not, "How much can I afford to lose?", but rather, "How much can I afford to pay for winning?" The carrier who wins bid after bid, but watches his profits slip away because of the insufficiency (from his own perspective) of his "successful" bids, has been seduced by auction mentality into failing to ask himself this key question.

Any competitive bidding opportunity, thus, should be analyzed first on the basis of the specifications of the RFP—those factors the shipper has declared to be immutable. The specifications should be compared carefully to the carrier's own concept of where he is now and where he wants to go, in terms of his own marketing strategy.

First, the traffic itself: Does (do) the commodity(ies) fall within the general framework of what the carrier wants to haul? In other words, does the carrier's marketing strategy contemplate movement of this type of traffic? If not, the place for his copy of the RFP is in the round file near the manager's feet, not on his desk.

Next, what about the geography? Will the traffic complement the carrier's existing operations, and/or jibe with his plans for market expansion? Again, if the answer is no there's no need to explore the matter further; this is a bid the carrier doesn't *want* to "win."

In a significant number of instances the bidding competition is decided right here. That is, a particular carrier will find the traffic so extraordinarily attractive, within the context of his other operations, that he will be in a position to submit a bid that no rationally motivated competitor can match. For example, the RFP may offer traffic moving from point A to point B at just the time when a particular carrier has a problematic imbalance of traffic in the opposite direction; for this carrier, that traffic is so alluring that he's willing to submit a bid far below his own or any competitor's fully allocated cost level.

Next, the carrier should carefully scrutinize the details of the service requirements of the RFP. Again, he should consider them within the context of his own capabilities and marketing objectives; and only if the two are more or less in consonance should he proceed. A trucker aiming at point-to-point truckload movements, for example, should avoid invitations to bid on peddle-run service, and so on. Once more, the carrier's first priority should be to implement his own marketing strategy, not grab indiscriminately at revenue opportunities that force him along paths at odds with what his strategy contemplates.

Here, however, the carrier should not be overly rigid about adherence to his predetermined marketing plans. No strategy, no matter how well conceived, can take into consideration every possible contingency; and no strategic planner should regard himself as so omniscient that every modest deviation from his planning must be rejected out of hand. At least the possibility of amending the existing plan should be given consideration before the RFP is discarded.

Where the service requirements spelled out in the RFP diverge radically from the standards under which the carrier now does and/or wishes to operate, there will likely be no room for compromise. If motor carrier transit-time specifications are so tight that they would require tandem operations and the carrier's fleet includes no sleeper cabs, for example, that carrier should probably not bid. If refrigerated equipment is needed and the carrier has none, again no bid is indicated.

But if the divergence is less drastic—if there's a more modest gap between the service the shipper wants and what the carrier wants to provide—there are options other than simply throwing the RFP into the wastebasket. First, the carrier can reconsider his own planning; perhaps it might be worthwhile offering this or that additional service. Certainly such reconsideration should not be of a casual nature; an appropriate revision of the carrier's overall strategic planning to accommodate the new ideas is called for. But such plans are not, after all, etched in stone, and the prospect of the traffic offered by the RFP may warrant a second look.

Alternatively, the carrier may elect to invite the shipper to reconsider *his* decisions, by submitting a bid at variance with the RFP specifications in certain particulars.

This is an approach that has to be handled very gingerly

indeed. To begin with, the carrier is obviously, by deviating from the RFP specifications, in effect telling the shipper he doesn't know what he wants. This is bound to have a negative impact on the shipper manager who prepared the RFP, and probably also on the shipper's organization as a whole. Moreover, many shippers will routinely discard bids that don't exactly match their pre-established standards, so the likelihood of success isn't high.

On the other hand, where the specification(s) involved isn't (aren't) central to the RFP—where the variance is in a secondary area—a low bid will probably attract notice. For an obvious example, if the shipper has asked for full carrier liability and the carrier submits a bid that calls for limited liability but, at the same time, is 10% lower than the next-low bid, the shipper will probably be impelled to at least look at the alternative of buying separate insurance for his shipments. And if the insurance cost plus the carrier's bid still works out to be lower-cost than the lowest full-liability bid. . . . Well, certainly the shipper has some thinking to do.

The point is that shippers are not bound, legally or otherwise (*see below),* to the precise specifications of their RFP's. It is rare indeed that shippers will accept bids that in any major way don't coincide with the RFP specifications. Small-scale variances, on the other hand, may well be attractive if accompanying a bid that's significantly lower than anyone else's. A carrier would be foolish to count on winning very many bidding competitions with service offerings that don't comply strictly with the RFP to which he's responding; but there are times when judicious revision of the RFP standards may be appropriate.

By the same token, carriers should be alert to RFP's that don't appear to adequately describe the service the shipper clearly wants. If the RFP calls for palletized shipments that are to be loaded and unloaded manually, for example, something is pretty clearly amiss. If no transit-time standards are specified on shipments of goods the carrier knows, from independent sources, to be extremely hot items on the market, the likelihood is that this was an oversight. In such circumstances, one alternative is for the carrier to draw the omission or conflict to the shipper's attention so that he can amend the RFP; but another, sometimes more astute (in marketing terms) one is to keep silent and submit a bid that corrects the apparent mistake.

Another such approach is for the carrier to submit a partial bid—a bid covering only a portion of the traffic involved. In some instances, as already discussed, shippers will actively encourage partial bids; but even when they do not, a partial bid may, once again, attract notice if it is substantially lower than bids of other carriers respecting the traffic in question.

In partial bidding, however, the carrier must be alert to circumstances where the shipper has "bundled" both desirable and undesirable traffic in order to ensure a reasonable rate/service package for the latter. If it appears that the shipper is intentionally grouping its traffic to avoid such problems, there is little likelihood that a partial bid which carefully "creams" the traffic will be accepted. On the other hand, if there appears a reasonable prospect that the unbid traffic will prove attractive, by itself, to other carriers, a partial bid may excite the shipper's interest even where it was not nominally permitted under terms of the RFP.

No matter what the circumstances, if the carrier's bid deviates in any way at all from the RFP specifications this fact should be strongly emphasized, both in the bid itself (in whatever format is required) and in a cover letter accompanying it. The carrier should make every effort to draw the shipper's attention to the fact that his bid is based on factors other than what the RFP described. There is simply no point to conducting further dealings under mutual misunderstandings; moreover, when the deviation later becomes apparent, the shipper may feel he was deceived by the carrier, which bodes poorly for any future relationship.

As a general rule, of course, it is advisable to stick closely to the RFP's list of specifications—that is, to offer precisely the service (no more, no less) the shipper says he wants. The bid elements of the RFP are, of course, another matter.

Here the carrier will want to consider the relative priorities the shipper has placed on the different factors on which bids will be evaluated—provided, of course, the shipper has spelled them out.* Major emphasis should be placed on the one or (at most) two elements deemed in the context of the RFP to be of greatest importance. The only exception to this rule is where the carrier believes his greatest strength lies in an area rated as secondary by the RFP, *and* he is fairly certain he has a significant edge over the competition in that area. In such circumstances a truly spectacular showing on the secondary area can bring success to a bid that, in areas nominally deemed of greater importance, is not No. 1.

Again, however, the carrier should not count on success by such means. For the most part shippers will insist on evaluating bids in accordance with the criteria they have specified; and a carrier that does excellently in secondary areas but falls down somewhat on the primary criterion is most likely going to wind up an also-ran.

RFP demands for particular bidding formats, auxiliary documentation and the like should always be observed scrupulously; and so should bidding deadlines. In some cases shippers will stand still for deviations from the RFP in any and/or all of these areas; but it's obvious that, all other things being equal, bids that don't meet the technical requirements are not likely to be the ones selected. Moreover, nothing can be more frustrating than to submit what would have been the low bid only to have it disqualified because it wasn't submitted in the required form, it arrived a few minutes late, etc.

As discussed above, for the most part a carrier should base his decisions to submit bids, and the bids he submits, primarily on factors concerning his own operations. Does he want the traffic? Can he meet the service specifications—and if so, at what price

---

*If the shipper has failed to indicate exactly on what basis he will evaluate bids, price—that is, line-haul rates—should always be given primary attention. For the most part competitive bidding is a price-oriented strategy for shippers; in most companies the Traffic Department's effectiveness is measured principally or exclusively on the basis of economics, and the carrier should (absent any indications in the RFP to the contrary) take this into consideration in submitting his bid.

(while still earning a profit)? Bidding blindly against unknown competitors, he really has little other rational option.

At the same time, he should certainly not simply ignore what he knows to be the realities of the marketplace. He may have no precise idea whom he's competing against in this particular instance; but he certainly does (or should) have a pretty good notion of what competitors he faces in any given sector of his chosen market, and how they are likely to respond to the particular RFP being considered. And obviously his bid should be prepared with these known competitive factors strongly in mind.

The calibre of competition he expects to face, and his estimates of the other bids with which he must compete, should influence not only the level of bid submitted by the carrier but also whether he decides to bid at all. Realistically speaking, competitive bidding is not, and should not be regarded as, a game played for its own sake; any rational person should have no difficulty recognizing this. Human psychology, however, isn't always grounded in 100% realism.

Thus, a carrier who submits bid after bid notwithstanding his expectation that he will be underbid by the competition may thereby earn for himself the label of "loser" in the eyes of shippers. That can impair his ability to secure other traffic that is not subject to competitive bidding, and can also place him in an awkward position if and when he *does* submit the low bid. If a carrier whose bid has been high 10 or 20 times running suddenly appears as the low bidder this time out, the shipper can be excused if he decides either to re-bid the traffic (on the premise that if a chronic loser's bid is low, he will probably do better with a second effort) or to award the traffic to another, higher bidder (because he no longer accepts this particular carrier as being "for real").

To avoid being stigmatized as a loser, a carrier should never bid on traffic he doesn't feel he has at least a reasonable chance of winning. Perhaps indeed, as Alfred, Lord Tennyson wrote, "'Tis better to have loved and lost / Than never to have loved at all"; but in the world of business, as distinguished from that of romance, it's preferable to give up long-shot opportunities than submit too many losing bids in search of them.

Nor should the carrier either take, or seem (in the eyes of shippers) to take, losing lightly. Efforts should always be made to follow up on any bids submitted, inquiring of the shipper to whom the traffic was awarded and at what bid. To start with, this tells the shipper his traffic was of importance to the carrier, which may make a favorable impression that could have useful future implications. Secondly, any information the carrier can thereby glean may help him improve his bidding technique next time around. And third, it can uncover improprieties in the bidding process.*

Legal and Ethical Considerations

Under the law, the shipper is not bound by the terms of the RFP.

This will come as a surprise to many people, whose primary exposure to competitive bidding is through news media accounts of government procurement activities. With fairly dismaying frequency, scandals keep cropping up about how this or that governmental agency has failed to use the process properly and, as a result, is paying far more than pure competitive bidding would have required.

As described above, however, government is in a somewhat different position as to its procurement than are most organizations in the private sector. In many instances competitive bidding for governmental bodies is mandated by the law and/or regulations, so that failure to abide strictly by the process is illegal. And even where legality isn't in issue, political con-

*In one actual case, a non-winning carrier in a governmental bidding competition reaped an unexpected reward from this kind of follow-up. The government agency involved declined (as most shippers will) to answer the carrier's questions about what the low bid was; but the carrier was in a position to follow through with a demand for that information under "freedom of information" statutes, and did so. To both it and the agency's consternation, it discovered that its bid was in fact lower than the winning one; there had been a mistake in the evaluation process and, after pointing this out, the award was reconsidered and the traffic given to the inquisitive carrier.

siderations still make a failure to adhere to strict bidding standards an appropriate matter for public concern.

Things are otherwise for private-sector purchasers who elect to use competitive bidding strictly for its economic benefits. First, for the most part they do so voluntarily, without either statutory or regulatory pressure. Second, if they fail to stick by the terms of their RFP's their status as private enterprises makes them fairly invulnerable to public criticism.

In law, the RFP amounts to what is known as a "unilateral contract," in which one party makes a fairly open-ended offer that it will give thus-and-such to any other party who's willing to give or do the particular thing that is required in exchange. An offer of a reward for the return of stolen property falls under this heading; so does an advertisement to sell goods at a particular price.

And such offers, as with any form of unilateral contract (including the RFP), can be withdrawn or changed pretty much at the whim of the offering party. Rewards can be cancelled, sales prices increased, etc., without notice or legal penalty. This holds true right up to the time when the exchange of goods and/or services actually commences (or a binding *bi*lateral contract is executed)—which certainly hasn't taken place while the competitive bidding is still in progress.

In some circumstances the freedom to back away from an RFP can be of tremendous benefit to the shipper. Suppose, for example, he has issued the RFP prematurely, only to find belatedly that the document was seriously in error, and/or that circumstances unexpectedly changed after it was issued. Or suppose he is dissatisfied with all the bids he has received, and thinks he can do better by means of some other negotiating form. Other situations might also be postulated in which the shipper would not want to abide by the terms of the RFP; and he's under no legal obligation to do so.

Indeed, this holds pretty well true for all aspects of the process. That is, even though he may have stated he will take the low bid, the shipper need not actually do so; if he chooses he can change the parameters of the service he wants after all bids have been submitted; he can decide unilaterally whether to consider bids submitted late, or not using the proper format, etc. He can do, in sum, pretty much as he likes without reference to what he said in his RFP.

As a practical matter, however, shippers will only rarely take major advantage of this legal freedom. To begin with, a shipper who has gone to the trouble of putting out an RFP has invested considerable time and energy into preparing it, and for that reason alone is unlikely to repudiate it except for good reason. Secondly, few shippers can be unaware of the consequences, in terms of both marketplace reputation and vendor relations, of going back on their word (as it were)—consequences that, for publicly held companies, can also extend to stockholders and/or investors. Third, unjustified rescission of an RFP tends to offend the ethical sensibilities of corporate managers, both as individuals and in their capacities within their companies.

Regrettably, however, not all corporate managers maintain the highest of ethical standards—and therein lies the greatest potential for abuse of the competitive bidding process.

Probably the most common type of abuse is where competitive bidding is employed mainly as a device to put pressure on existing suppliers. That is, the shipper ostensibly opens up a particular increment of traffic for bids, indicating he will award the business to the lowest bidder; but privately he advises one or more favored carriers that it or they need not participate in the bidding, but instead will be given an opportunity to match the low bid at a later time. In actuality, that is, bids are solicited only to provide negotiating leverage for the shipper's dealings with his preferred carrier(s).

Such bid "rigging" is not exceptionally common; but it does occur, and sometimes in more virulent forms. Cases of out-and-out bribery of industrial traffic officials and the like occasionally surface, so that in some instances it will be individual managers who profit most from the competitive bidding process, not the shipper itself.

Because of the non-binding nature of the RFP, carriers have little or no legal recourse in most such cases. There are, to be sure, statutes outlawing commercial bribery; and in some states carriers have a right to sue if they discover that their supposedly secret bids have been indiscreetly revealed to their competitors. For the most part, however, the carriers' recourse in such cases is economic; they may (and should) simply decline to participate in competitive bidding processes whose integrity they deem suspect. And they are certainly free to "pass the word" when they learn that a particular shipper has been using the process in such unethical fashion (although, under the laws of slander and libel, they had better make very sure of their facts before doing so).

Although the shipper's RFP is not, in itself, binding, the same is not always true of the bids submitted by carriers. The circumstances will vary according to the manner and form in which bids are submitted; and the carrier does have complete legal freedom to change, revise or even withdraw a bid at any time prior to being notified by the shipper that the bid has been accepted (even if the deadline for bidding has passed).

Once accepted by the shipper, however, the bid may be irrevocable by the carrier where contract carriage is involved. That is, the carrier's submission of the bid and the shipper's acceptance of it may be deemed legally sufficient to establish the existence of a bilateral contract, to which both sides are now committed. The situation here is a good deal murkier with regard to common carrier bidding, especially where regulatory law requires tariff publication by the carrier before any rates may go into effect. However, deregulation has engendered, both in the courts and at the ICC, some hints that carriers may today be held to non-tariff agreements in some circumstances; and at the least, the law should be considered in a state of flux on this issue.*

As a practical matter, of course, the question will almost certainly not arise. A carrier who wishes to withdraw his bid even after notification of acceptance by the shipper will in all likelihood be allowed to do so; few shippers are eager to do business with carriers who don't want to do business with them. Again, the

---

*Although this statement must, again, be read with the implicit modifier, "as of the time this was written," it appeared improbable that there would be any major clarification of this uncertainty for at least the foreseeable future.

most likely sanctions to be applied against carriers in these circumstances are economic; a carrier who backs off a formally submitted bid after it has been accepted had better have (and be able to communicate to the shipper) an extraordinarily good reason if he ever expects to be regarded with any favor by that shipper in the future.

In sum, economics, not the law, is the primary vehicle to hold both shippers and carriers in line with regard to ethics. A single ethical failure can prove almost permanently disabling to a shipper's or carrier's ability to develop future relationships; and for that reason such practices should be assiduously avoided.

The Future of Competitive Bidding

To what extent will competitive bidding continue to play a part in the transportation industry's future?

As of when this was written, the question appeared imponderable. The industry, in both shipper and carrier quarters, was still reeling from the effects of the rapid-fire legislative and administrative deregulation that had been thrust upon it, and the form(s) that future shipper-carrier relations will take were far from being clearly established.

Prior to about 1980 or so, competitive bidding was all but unknown in transportation. It was introduced into the industry by major shippers who were looking for the fastest possible means of gaining the benefits of deregulation-spurred competition. And it has served that purpose quite well—well enough that it seems likely to remain in at least some use for the foreseeable future.

The choice, ultimately, lies with shippers; as the buyers of transportation services, they have the power to define the manner in which they will do their purchasing. And even though some carriers still adamantly decline to participate in bidding competitions, the industry is so heavily populated (especially in the motor sector, where competitive bidding is being most widely used) that it seems unlikely enough carriers will boycott the process to enforce that view on the marketplace.

Competitive bidding has both its strengths and its weaknesses. On the plus side, it allows shippers to conduct a form of negotiations with many carriers at the same time, selecting the best deal available. It also encourages carriers to hold rates down, since they know at the time they make their bids that, to be successful, they must underbid many competitors.

But there is a down-side, too. First, there is the difficulty of preparing the RFP with sufficient precision and completeness to eliminate potentially costly slip-ups. Second, use of competitive bidding tends to signal carriers that the shipper is primarily a "price-only" customer (price being, of course, the most usual bid element)—that is, that he will be satisfied with a minimum level of service within the range of the RFP specifications (which is, as a result, the best he can normally expect). And third, many carriers, especially in the trucking sector, are still not used to competitive bidding, and decline as a matter of policy (or inexperience) to respond; so the shipper is cutting off some potential suppliers simply by the manner in which he approaches the marketplace.

Within these parameters, it is obvious that the competitive bidding tool must be used judiciously. Where it is appropriate, it is a powerful addition to the shipper's negotiating arsenal; where it is not, its use can cause a lot more problems than it solves.

10

ANTITRUST LAW AND RATE NEGOTIATIONS

Just as deregulation has eliminated many of the major legalistic limitations on the shipper-carrier relationship, so has it added one relatively new to those in the transportation industry—the rigors of antitrust law.

For the most part, both in the technicalities of the law and by custom, the transportation industry has considered itself largely immune from the antitrust standards so long applicable to other economic sectors. Today, though, not only the statutes but, even more, the recent activities of the Justice Department, the Federal Trade Commission, the ICC and the courts have made it clear that antitrust law has a new meaning and application in transportation.

There are three fundamental guidelines for carriers and shippers seeking to avoid problems in this area:

(1) The company—carrier or shipper—should conduct its business in compliance with the antitrust laws.

(2) The company should *appear* to be conducting its business in compliance with the antitrust laws.

(3) The company should be able to recognize incipient antitrust problems so as to seek timely legal advice.

In other words—be innocent. . . *look* innocent. . . and seek legal advice when you're not certain you meet both criteria. The problem is that none of these things will be easy for the transportation manager who lacks experience or background in the very broad field of antitrust law.

The basic objective of antitrust statutes (principally the Sherman and Clayton acts) is to ensure, insofar as possible, "pure" economic competition in the supply-and-demand marketplace.

Not surprisingly, it is in the self-interest of almost any individual or firm engaged in business to skew the supply-and-demand balance in his or its favor. The most common ways by which this may be achieved are by (1) conspiring with others to upset or distort that balance, or (2) seeking to establish some form of marketplace dominion by any of various means.

Antitrust statutes outlaw both. Indeed, they go further; they bar *any* practices and activities that may, or (and this is important) *are intended to,* have significant "anticompetitive" effects of any sort. The statutes empower the courts to give orders to prevent such anticompetitive results from ever occurring, or, if they have already occurred, to (insofar as possible) undo the damage.

It should be emphasized that these statutes are construed very liberally by both the governmental bodies responsible for enforcing them (primarily Justice and the FTC) and the courts who are responsible for their interpretation and application. As business alters its approaches to gain marketplace advantages, the law is thus able to follow into new areas and bar practices that may never have been previously adjudicated, and weren't even contemplated by Congress when the governing statutes were enacted. Thus, there will not always be clear guidelines based on past readings of the law that will assist business in complying with it in the future—especially where transportation, to which many antitrust applications are quite new, is concerned.

Of particular importance, antitrust law applies to both seller *and* buyer. Thus, shippers as well as carriers may be adjudged guilty of violating these provisions of the law. And such violations, even if unwitting, can be disastrous for those who are found guilty—or sometimes for those accused and found *not* guilty.

First, the legal costs of mounting a defense to an antitrust accusation can be horrendous. Because of the complexities of the law and its specialized nature, those costs—in terms of both attorneys' fees and development of the necessary evidence—can and do rapidly add up to hundreds of thousands of dollars or more. Thus, many companies innocent of any wrongdoing will elect to settle a case (and pay the heavy fines and/or penalties involved—*see below)* rather than bear the expense of contesting the accusation.

Second, fines and civil penalties run high for violations. A million dollars is not uncommon, and in some instances violators have paid still higher sums.

And third, and perhaps most frightening, any private party (a competitor, a customer, etc.) who claims economic injury as a result of antitrust violations is *automatically* entitled to *treble* damages (monetary awards equal to three times the amount of economic damages proved) if he prevails in court. Even if the violation was wholly unintentional, completely without any nefarious intent, the law leaves courts no option to reduce the awards in such cases.

Until recently the transportation industry was largely shielded from application of antitrust law by virtue (directly or indirectly) of its regulated status. There were four separate sheltering mechanisms—but all have either been eliminated altogether or substantially reduced by deregulation.

• Statutory immunity—specifically, the immunity extended to the "collective ratemaking" activities of rate bureaus under *the Reed-Bulwinkle Act* incorporated into the Interstate Commerce Act in 1948. Collective ratemaking is, of course, a euphemism for what in other economic sectors is rigorously prohibited as "price-fixing." This immunity no longer exists for airlines (which had to secure a special antitrust exemption in 1984 merely to discuss rescheduling, for the purpose of easing airport congestion, with one another). ICC-regulated railroads, motor and water carriers and freight forwarders retain some shreds of their prior immunity, but the bureaus are so constrained in what they may do that their role as pace-setters in transportation ratemaking is clearly past. In addition, some recent court decisions have hinted that antitrust protection for carriers who engage in collective ratemaking may be quite limited even where the law nominally allows bureau action (*see below).*

• The so-called *Noerr-Pennington doctrine* (named after two Supreme Court cases in which it was enunciated), which protects businesses if they collaborate for such purposes as legislative lobbying, court litigation and—of special importance to transportation—actions before government agencies. It is not even required that such activities actually take place, so long as (1) there is the legal right to take action, and (2) talks are limited to that area. Deregulation, by removing the right of shippers and carriers to seek regulatory relief in many areas, has curtailed this protection substantially.

• The *Keogh doctrine* (likewise named after the Supreme Court case wherein it was first spelled out), which relieves regulated transportation carriers of potential treble-damages antitrust lawsuits based on rates, charges, rules, etc., published in tariff form. The reasoning behind this was that carriers' shippers and competitors already had the right to seek regulatory relief from tariff provisions injurious to them, and that to give them the potential for antitrust relief as well would be both redundant and a form of "double jeopardy" (of sorts) for the carriers. Deregulation of the airlines, and expanded regulatory exemptions for other modes (especially rail), as well as the sharp growth of contract carriage, have all served to substantially reduce the volume of transportation performed under government-filed tariffs, and, accordingly, the scope of this immunity.

• The largely informal, but widely observed, *"regulated industry protocol"* in Washington's bureaucracy. Essentially, this obliges antitrust enforcers such as the Justice Department and the FTC to take a "hands-off" attitude toward industries regulated by other Federal agencies; the FTC Act actually incorporates a specific provision to this effect applicable to that agency, and Justice, while not constrained by statute, has also historically observed it as a general (though occasionally violated) policy. Deregulation has opened many doors here; and even where it retains regulatory jurisdiction, the post-1980 ICC has been encouraging greater Justice Department antitrust oversight of transportation subject to its (the ICC's) jurisdiction.

Indeed, transportation has become something of a "target industry" for antitrust enforcers.

In the newly active (in transportation) antitrust arena, certain things must be approached with considerable caution. Most particularly does this apply to rate bureaus, on which special antitrust attention has been focused.

Most antitrust enforcement and litigation concerning rate bureaus have focused primarily on the bureaus' carrier members who, after all, are the prime movers in these forums. But shippers, too, can find themselves caught up in broad antitrust sweeps if they are unlucky—especially since the prevailing legal view is that any antitrust immunity extended to the bureau applies only to carriers, and not at all to shippers. Consider, for example, the potential for problems if a shipper successfully negotiates particular rates through a bureau, but its competitor is unable to do the same—or, worse, if several shippers collectively negotiate favorable rates withheld from other competitors.

Some Recent Cases

Four recent cases indicate how strongly antitrust standards are being applied in connection with rate bureau activities—and indicate, too, that even the vaunted bureau immunity (to the very limited extent it still exists) should not be regarded as nearly so comprehensive as it was once thought to be.

The so-called "Lake Erie coal docks case" was a landmark in this regard. As background, substantial volumes of coal move from mine areas in upper Michigan and Minnesota via Great Lakes barge carriers to Ohio's Lake Erie ports, for further movement beyond to inland destinations served from those ports. The following is a contemporary summary of the main allegations in the antitrust indictment handed down by a federal grand jury in October, 1981, against five railroads—the Baltimore & Ohio, Bessemer & Lake Erie, Chesapeake & Ohio, Conrail (as successor-in-interest to Penn Central, which allegedly committed the violations) and Norfolk & Western:

"The carriers are accused of trying [by means of provisions of rate-bureau tariffs] to monopolize coal traffic from Lake Erie ports to steel mills in Ohio, Pennsylvania, West Virginia and Kentucky by:

"• restricting line-haul commodity rates to traffic moving from railroad-owned, but not private, docks;

"• refusing to lease dock space to at least one shipper (Litton Industries); and

"• establishing 'arbitrary' charges on truck pickups of coal from rail docks.

"I'm* not, of course, privy to these railroads' thinking as to this situation. But I can envision a thought process that might account for the circumstances described in the indictment, which goes something like this:

"'I, a railroad, have spent a lot of capital building docks in order to develop line-haul business out of those docks. I'm not interested in letting others—especially truckers, my arch-competitors—reap the economic benefits (in terms of line-haul revenues) of my investment. I'm also not interested in encouraging competing private docks, which might detract from that investment by siphoning traffic away from my docks. So in these cases my self-interest dictates that I be fairly uncooperative.

"'On the other hand, the nature of railroading being what it is, I can't expect to handle all the line-haul out of my docks; I don't serve all the destinations. So it *is* in my interest to cooperate, through the rate bureaus, with my fellow railroads, because they will then cooperate with me.'"

It should be emphasized that the actions cited in the indictment in this case took place in the early 1970's, before first the Railroad Revitalization and Regulatory Reform ("Four-R") Act and later the Staggers Rail Act imposed any regulatory limitations on the antitrust immunity of the rail rate bureaus. The carriers, that is, had every reason to believe their actions, no

*The quotation is extracted from the November 2, 1981, *Barrett Transportation Newsletter;* the "I" is the *Newsletter's* editor, who is also the author of this book.

matter how motivated, were sheltered from antitrust law by both the Reed-Bulwinkle Act and the Keogh doctrine.

They were wrong.

After the antitrust indictment was handed down the carriers confidently sought from the Federal district court hearing the case a "summary judgment" dismissing it. To their astonishment, the court refused (in a decision later upheld on appeal), holding that rate bureau immunity did not extend to actions taken with "anticompetitive intent." In other words, it was the reasons that underlay the carriers' actions, not simply the actions themselves, on which they would be judged.

Four of the five carriers chose not to carry their defense further in the face of this ruling; B&O, B&LE and C&O all agreed to pay forfeitures of $1 million each, and Conrail paid $100,000 (its penalty was lower because it was actually Penn Central that allegedly committed the violations, not Conrail itself). N&W defended itself and won acquittal on the ground that there wasn't enough evidence proving it was part of any antitrust-violating conspiracy; but in its decision the court made plain its opinion that (1) such a conspiracy did, in fact, exist, and (2) it was unshielded by the bureau's antitrust immunity.

The second and third cases, brought respectively by Clipper Exxpress against the Rocky Mountain Motor Tariff Bureau and Laker Skytrain against airlines setting rates "collectively" under the aegis of the International Air Transport Association, hinged on the same basic issue of "anticompetitive intent." Both carriers accused their bureau-based competitors of trying to discourage their (Clipper's and Laker's) innovative price competition by means of nominally antitrust-immune bureau actions; in both cases courts held, in preliminary rulings, that the bureaus' immunity did not extent to actions taken with anticompetitive intent. And both were ultimately settled out of court by the defendants paying very substantial sums to the two plaintiff carriers.

These cases, at least, hinged on the intent underlying *actions* taken by the defendant carriers; the fourth revolved around the intent behind mere *words.* The Federal Trade Commission instituted action against an intrastate truck rate bureau—the Massachusetts Furniture and Piano Movers Association (MFPMA)—on the novel ground that, by merely advising its members of the rate proposals of individual carriers, the MFPMA was implicitly urging other carriers to take identical ratemaking actions of their own. Once again the case was settled out of court, this time by means of a "consent agreement" by the association that it would cease communicating such rate information to its membership.

It should be emphasized that none of these cases (nor any other similar ones) went all the way through the litigatory process; none, thus, stands as clear-cut legal precedent to support future legal rulings.* Even so, these cases, taken together, should be deemed a shot across the transportation industry's bow—a dramatic warning that bureau participation that involves cooperation with competitors on any rate matters, no matter how seemingly antitrust-protected, is fraught with peril.

Indeed, the warning about rate bureau activity should realistically be extended to virtually all forms of collective action. Major industry organizations have gone to extraordinary lengths to protect themselves and their members from antitrust problems, even to the extent of having experienced antitrust attorneys on hand at major meetings. Smaller groups, however, lack such expert advice, and it is not difficult for talk to stray into forbidden territory. In this context it is important to note that ignorance of the law will never be regarded as an acceptable excuse in an antitrust action—nor will purity of heart or other pleas that no violation was intended.

The case of American Airlines President Robert Crandall stands as a clear warning that ill-chosen words, even in casual

---

*This *caveat* does not apply to the "anticompetitive-intent" rulings which, now having been handed down by at least three separate courts and supported on appeal in the Lake Erie case, may indeed be regarded as reasonably precedent-setting. In addition, the Lake Erie case has led to the filing of at least two private-party suits (by Litton and an agent representing motor carriers) which were still pending as of the time this was written; the decisions in these cases, if they go through trial, may shed further light on this question.

conversation, can pose severe antitrust risks. Crandall was recorded, by means of a Justice Department telephone tap, as suggesting to a competitor (Braniff, prior to that carrier's bankruptcy) that both increase fares. Nothing came of the suggestion, which was little more than an off-the-cuff remark; but Justice indicted Crandall merely on the ground of his suggestion, and the courts agreed that this is a proper ground for trial. Once more, the case was settled out of court (by means of a fairly innocuous consent agreement); but it still stands as a clear proof that business managers must watch their words at *all* times.

Other Facets of Antitrust Law

It may happen that even the most careful of managers will sometime find himself at a gathering where the discussion turns to rates or services in a fashion that seems dangerously specific. In these circumstances, antitrust experts are unanimous in their advice: The manager should leave immediately—and should also make certain others present notice his or her departure so that, if antitrust problems later arise, there will be witnesses to the fact that he or she was not there.

Some caution must also be used in connection with shipper associations. The law explicitly allows buyers, in any industry, to pool their purchasing power for purposes of negotiating and/or consummating deals with vendors. But this immunity holds good only so long as the activities of such associations are limited strictly to this purpose, *and* do not serve to distort or skew marketplace competition.

Potential danger areas here include associations with an overly heavy concentration in a particular industry (where it might be felt that the members were conspiring through the association to parcel out market shares in some agreed fashion, rather than competing with one another in the area of transportation-cost reduction), and associations that exclude some shippers from membership (thereby distributing membership benefits unequally, and perhaps, in antitrust enforcers' eyes, non-competitively). Shipper associations that employ contract carriers must be particularly wary; the ICC has advised such associations to check out their contracts with the Justice Department's Antitrust Division.*

Obviously, the safest way to avoid running afoul of antitrust law is to negotiate transportation arrangements strictly one-on-one. That means one shipper, one carrier, and nobody else in the room. And even then, both shipper and carrier are probably best advised not to discuss their negotiations with others too freely. Even though, as to regulated common carriage, the tariff-publishing requirement makes details of the shipper-carrier relationship pretty much a matter of public record, premature disclosure of agreements before tariffs are published—or excessive talk about unregulated agreements—could be construed as illegally tipping others off about competitive matters.

The Justice Department is acutely sensitive to such "tip-offs." In the mid-1970's, for example, it brought suit against several paint manufacturers for allegedly letting one another know in advance about pricing changes. The manufacturers had simply been engaging in the not-uncommon practice of sending out press releases to the business media, and letters to major customers, a few days in advance of the price hikes. But they were forced to halt this practice by the antitrust action.

A final area in which the transportation manager should beware, from an antitrust standpoint, involves so-called "tying arrangements." This phrasing is used to describe the managerial decision of a buyer or seller to "tie" two otherwise unrelated transactions together into a single package. If that decision is forced on those with whom the buyer or seller is doing business, an antitrust violation can result. For example, this might be deemed to preclude an automotive tire manufacturer from requiring that trucks hauling its freight be equipped with its tires, a stationer from insisting the the carriers he uses submit their freight bills on the paper in which he deals, etc.

---

*It must be noted, however, that even this will not always protect against antitrust prosecution; the Antitrust Division's opinions in such situations are informal and non-binding.

In this connection it is worth noting that many shippers view traffic in a unitary fashion for negotiating purposes. Nearly every shipper will have some traffic components that are more desirable, from the carriers' perspective, than others. To ensure service on their less-desirable traffic as well, they may require that any carrier serving them agree to handle it in order to participate in the (more desirable) balance of their traffic. Some carriers, similarly, will insist that they receive a share of a particular shipper's "cream" traffic before they will agree to handle his less desirable movements.

So far, at least, no antitrust challenge of these practices has been raised. However, a not-dissimilar case involving motion picture distributors may have ominous implications. The distributors were requiring theater chains to show unpopular films as a precondition to giving them access to lucrative "blockbusters." Antitrust litigation resulted in an end to this practice—which has, obviously, many parallels to the above-described transportation practices.

It is not the role of this book to offer any carrier or shipper detailed advice on antitrust-related matters. The author is not an antitrust lawyer, and does not pretend to expertise in this highly complex area of the law. When questions of antitrust law arise, an attorney should always be consulted—an attorney, moreover, who specializes in antitrust practice.

Finally, the most important rule of all in connection with antitrust law is—When in doubt, don't. If a particular action appears in any way questionable from an antitrust standpoint, the manager should avoid it. Even if it does not actually violate the law, the mere fact that it *seems* to do so may be enough to provoke damaging and costly litigation.

Index

J

K

L

M

N

O

U

V

Z